Detection and Diagnostics of Plant Pathogens

Plant Pathology in the 21st Century

Volume 5

For further volumes:
http://www.springer.com/series/8169

Maria Lodovica Gullino • Peter J.M. Bonants
Editors

Detection and Diagnostics of Plant Pathogens

Springer

Editors
Maria Lodovica Gullino
AGROINNOVA
University of Torino
Grugliasco, Torino, Italy

Peter J.M. Bonants
Plant Research International,
Wageningen UR
Wageningen, The Netherlands

ISBN 978-94-024-0125-7 ISBN 978-94-017-9020-8 (eBook)
DOI 10.1007/978-94-017-9020-8
Springer Dordrecht Heidelberg New York London

Printed on acid-free paper

Springer is part of Springer Science+Business Media (www.springer.com)

Foreword

This volume continues the series of books on "Plant Pathology in the 21st Century", which started in 2010, in cooperation with the International Society for Plant Pathology and contains the lectures given at the 10th International Congress of Plant Pathology (ICPP 2013) held in Beijing, August 25–30, 2013.

At such Congress, several sessions dealt with aspects of detection and diagnosis of plant pathogens, which represent two fundamental steps in disease management decisions.

For both detection and diagnosis, new tools and technologies have been developed, which are often replacing old methodologies, permitting to be faster, more specific and more precise.

A quick and reliable detection method in combination with decision support systems is fundamental in order to reduce the damages caused by old and new pathogens, thus permitting to reduce the number of treatments and to contain the potential losses.

Molecular methods are available for a number of pathogens and the volume provide good examples of application in different production sectors. Innovative techniques and methods will be described to detect and identify different targets: destructive and non-destructive, air- or soil-borne, human and plant pathogens, in plants or seed-born, native or emerging pathogens, on-site or lab-based. All to support international organizations to secure global trade and agriculture all over the world.

We believe that, besides representing a written testimony of ICPP 2013, this book will be useful for all plant pathologists as well as students in advanced courses.

We wish to thank all the colleagues who accepted to be part of this book, Zuzana Bernhart and her group at Springer for their continuous support and Laura Castellani for her skilful technical assistance.

Maria Lodovica Gullino
Peter J.M. Bonants

Contents

Part III Role of Diagnostics in Plant Disease Management

Contributors

Brett J.R. Alexander Plant Health and Environment Laboratory, Ministry for Primary Industries, Auckland, New Zealand

Will Baldwin Department of Plant Pathology, Biosecurity Research Institute, Kansas State University, Manhattan, KS, USA

Peter J.M. Bonants Plant Research International, Wageningen UR, Wageningen, The Netherlands

Neil Boonham Plant Protection Programme, The Food and Environment Research Agency, York, UK

Thierry Candresse INRA and University of Bordeaux Virology Team, UMR 1332 Biologie du Fruit et Pathologie, Villenave d'Ornon Cedex, France

Joanne R. Chapman Previously Plant Health and Environment Laboratory, Ministry for Primary Industries, Auckland, New Zealand

Section for Zoonotic Ecology and Epidemiology, Linnaeus University, Kalmar, Sweden

Kerry R. Everett The New Zealand Institute for Plant and Food Research Ltd., Auckland, New Zealand

Jacqueline Fletcher Department of Entomology & Plant Pathology and National Institute for Microbial Forensics & Food and Agricultural Biosecurity, Stillwater, OK, USA

Angelo Garibaldi Center of Competence Agroinnova, University of Torino, Grugliasco, Italy

Giovanna Gilardi Center of Competence Agroinnova, University of Torino, Grugliasco, Italy

Maria Lodovica Gullino AGROINNOVA, University of Torino, Grugliasco, Torino, Italy

Steph Heard Department of Plant Biology and Crop Science, Rothamsted Research, Harpenden, UK

Michael J. Jeger Centre for Environmental Policy, Imperial College London, Ascot, UK

Jan Kreuze Laboratory of Virology, Peru International Potato Center (CIP), La Molina, Peru

Benedicte S.M. Lebas Plant Health and Environment Laboratory, Ministry for Primary Industries, Auckland, New Zealand

Lia W. Liefting Plant Health and Environment Laboratory, Ministry for Primary Industries, Auckland, New Zealand

Li Maria Ma Department of Entomology and Plant Pathology, National Institute for Microbial Forensics & Food and Agricultural Biosecurity, Oklahoma State University, Stillwater, OK, USA

Walter F. Mahaffee United States Department of Agriculture – Agricultural Research Service, Horticultural Crops Research Unit, Corvallis, OR, USA

Department of Botany and Plant Pathology, Oregon State University, Corvallis, OR, USA

Anne-Katrin Mahlein INRES – Phytomedicine, University of Bonn, Bonn, Germany

Charles Manceau Anses, French Agency for Food, Environmental and Occupational Health and Safety, Angers, France

Francisco M. Ochoa Corona Department of Entomology & Plant Pathology and National Institute for Microbial Forensics & Food and Agricultural Biosecurity, Stillwater, OK, USA

Erich-Christian Oerke INRES – Phytomedicine, University of Bonn, Bonn, Germany

Giuseppe Ortu Center of Competence Agroinnova, University of Torino, Grugliasco, Italy

Mark Payton Department of Statistics, Oklahoma State University, Stillwater, OK, USA

Megan K. Romberg Previously Plant Health and Environment Laboratory, Ministry for Primary Industries, Auckland, New Zealand

USDA-APHIS, Beltsville, MD, USA

Vittorio Rossi Università Cattolica del Sacro Cuore, Piacenza, Italy

James P. Stack Department of Plant Pathology, Biosecurity Research Institute, Kansas State University, Manhattan, KS, USA

Giuseppe Stancanelli European Food Safety Authority, Animal and Plant Health Unit, Parma, Italy

Ulrike Steiner INRES – Phytomedicine, University of Bonn, Bonn, Germany

Robert K. Taylor Plant Health and Environment Laboratory, Ministry for Primary Industries, Auckland, New Zealand

Jane E. Thomas Department of Plant Pathology, National Institute of Agricultural Botany (NIAB), Cambridge, UK

Gregor Urek Kmetijski inštitut Slovenije, Ljubljana, Slovenia

Joel L. Vanneste The New Zealand Institute for Plant and Food Research Ltd., Hamilton, New Zealand

Paul J. Verrier Department of IT and Computing, National Institute of Agricultural Botany (NIAB), Cambridge, UK

Lisa I. Ward Plant Health and Environment Laboratory, Ministry for Primary Industries, Auckland, New Zealand

Bevan S. Weir Landcare Research, Auckland, New Zealand

Jonathan S. West Department of Plant Biology and Crop Science, Rothamsted Research, Harpenden, UK

Guodong Zhang Center for Food Safety and Applied Nutrition, Food and Drug Administration, College Park, MD, USA

Part I
Technologies

New Developments in Identification and Quantification of Airborne Inoculum

Steph Heard and Jonathan S. West

Abstract Airborne spores initiate many fungal diseases of crops but can occur with a patchy spatial distribution or with a variable seasonal timing. New diagnostic methods are available for use on spores sampled in air to give a rapid and on-site warning of inoculum presence or to monitor changes in genetic traits of pathogen populations, such as the race structure or frequency of fungicide-resistance. Increasingly, diagnostic methods used on-site or even integrated with air sampling equipment are being developed. These include fluorescence and image analysis methods, DNA-based methods such as qPCR, isothermal DNA amplification (LAMP and recombinase polymerase amplification), antibody-based methods (fluorescence microscopy and resonance imaging, ELISA, lateral flow devices, and biosensors such as holographic or SRi sensors) and biomarker-based methods (such as detection of volatile or particulate toxins or other metabolites by electrochemical biosensor). By allowing a rapid detection, these methods can offer a direct warning of the presence of inoculum to direct disease control decisions. Air samplers are often used within crops, just above the crop canopy, or on aircraft (including UAVs) or on tall buildings. Their location affects the threshold of spore concentrations that translates to disease risk. The optimal deployment of air samplers varies according to how widespread the pathogen is, the type of air sampler used (particularly the rate of airflow sampled for volumetric devices) and the importance or value of the crop.

Keywords Optical sensing • Remote sensing • Biosensor • Inoculum detection • Immunological detection • DNA-based detection • Biomarker

---

S. Heard • J.S. West (✉)
Department of Plant Biology and Crop Science,
Rothamsted Research, Harpenden AL5 2JQ, UK
e-mail: jon.west@rothamsted.ac.uk

M.L. Gullino and P.J.M. Bonants (eds.), *Detection and Diagnostics of Plant Pathogens*, Plant Pathology in the 21st Century 5, DOI 10.1007/978-94-017-9020-8_1,

Introduction

Plant disease epidemics occur at different times and locations. Infection processes are directly influenced by the complexities of inoculum availability, growth stage of susceptible crop plants, and weather patterns (McCartney et al. 2006). Air-dispersed spores have been attributed to both long-distance introductions of pathogens over continental scales and local spread within or next to the field where the spores were produced (Brown and Hovmøller 2002; Gregory 1973). This can lead to small foci of infections when inoculum arrival is rare, uniform disease when inoculum is ubiquitous and disease gradients initiated from local areas of inoculum. Detection of airborne pathogen propagules that initiate disease is key information to drive decisions on crop protection measures and can also be used as an unbiased and easily obtained sample to monitor changes in genetic traits of pathogen populations.

Many airborne crop diseases are seasonal due to production of fruiting bodies only under certain conditions and spores are often released only after specific weather events. For example ascospores of many fungal plant pathogens are produced in fruiting bodies on crop debris in fields or field margins. These are often actively released after wetting by rain, which allows the spores to be ejected into more turbulent air to enhance dispersal and causes locally deposited spores to have perfect conditions for infection on wet leaves if the plant they have landed on is susceptible. In contrast, asexually-produced spores of rust fungi and powdery mildews are usually dry spores, passively released in windy conditions. This causes spore release to occur mainly during the day. Dispersal of airborne spores are often very local to the source but some spores are dispersed to altitudes over 10 km and have been documented to have dispersed over continental scales, often crossing oceans (Pady and Kapica 1955). Those that remain close to the ground may still be dispersed over 100 km, with spores within 100 m altitude often settling under gravity at night in relatively still air, falling at speeds between 0.03 cm s^{-1} to over 2 cm s^{-1} (Gregory 1973). Although spores of some species of fungi are very sensitive or not long-lived, others are able to tolerate UV light, desiccation and freezing (Gregory 1952). As a result, RT-PCR methods, which detects the more transient RNA in spores or culturing methods can be used to assess whether sampled spores are still viable (West et al. 2008). Even spores that are rain-splash dispersed may become aerosolised, particularly as a fine spray, caused by a combination of rain and strong winds. It has been reported that in Florida, the Asiatic citrus canker bacterium, *Xanthomonas axonopodis* pv. *citri* could be dispersed for up to 7 miles in severe rainstorms and tropical storms (Gottwald et al. 1992, 2001). Some fungal spores and bacteria are associated with ice-nucleation or enhanced condensation of water on their surfaces, leading to snow or rain formation, which returns the spores to ground from high altitudes (Fröhlich-Nowoisky et al. 2009; Morris et al. 2013). Recently various molecular methods suggest that bacteria are more prevalent in air than thought previously, when culturing methods were routinely used for assessment ($<$10 % of bacteria are able to be cultured in general media). For example, Morris et al. (2013) presented a highly plausible explanation that ice nucleating *Pseudomonas syringae* affected the hydrological cycle, by inducing rainfall, by ice nucleation activity to aid their dispersal, with isolates taken from a wide range of habitats (upland lakes, snow, soil, rivers, rock and plant leaves) each having a high likelihood of possessing effector genes necessary for plant infection. Plant pathogen populations often comprise many cultivar-specific races that are subject to selection according to the extent of cultivation of particular crop varieties. Van der Wouw et al. (2010) demonstrated how air sampling integrated with qPCR can be used for monitoring changes in race structure of *Leptosphaeria maculans*. Similarly, mutations that confer resistance to fungicides can be monitored for research and applied purposes (Fraaije et al. 2005).

Types of air sampler used in plant pathology are reviewed in Jackson and Bayliss (2011) and West et al. (2008). These can include passive samplers, which collect a deposit of spores and other particles onto an adhesive surface (such as a vaseline-coated microscope slide) and more usually, volumetric samplers, which include impactors such as the Hirst and rotating-arm trap, or cyclone traps, filters, impingers (which are necessary for efficient collection of very small particles) and virtual impactors (Cox and Wathes 1995; Lacey and West 2006). Additionally a type of electrostatic spore trap, the ionic spore trap has been produced (http://ionicsporetrap.com/). These have different pros and cons regarding air flow rate sampled, power consumption, sample period, collection efficiency, period they can be left unattended and ease of processing the samples. Impactors were originally intended for microscopy or culturing methods but some newer air samplers (miniature cyclone, MicroTitre Immuno Spore Trap, and the Ionic spore trap) have been, designed specifically for analysis using non-visual methods such as immunological and molecular diagnostics and most if not all older types are adaptable for these downstream diagnostic assays. Increasingly, samplers that sample into tubes or other vials are used to make processing steps user-friendly and even to facilitate automated testing of samples. Lab-based diagnostics applied to air samples suffer the disadvantage of the delay taken to transport samples to the lab but in some cases, this remains the only method available and may be more cost effective if multiple target organisms can be detected. However, a key factor to enhance this approach further, is the rapidly developing area of using new diagnostics, such as biosensors, lateral flow devices and isothemal DNA assays, that can detect pathogens rapidly and on site. For more generic studies of pathogens and the air-spora community, particularly for unknown fungi, bacteria and viruses or viroids in other particles, lab-based methods such as terminal restriction fragment length polymorphism (TRFLP), denaturing gradient gel electrophoresis (DGGE) and next generation sequencing are often used.

Direct airborne inoculum-based disease forecasting networks have been established for a few pathogens in a few countries and is of particular value if the disease incubation period is long e.g. in Poland for *Leptosphaeria maculans* (www.spec.edu.pl), which uses a network of Hirst spore traps and for vegetable brassica diseases in England (http://www.syngenta-crop.co.uk/brassica-alert/), which uses multivial cyclones. The latter collects a daily air sample at each site but these are only tested manually, using an on-site immunological test, if the weather data collected at the same site is conducive for infection. Direct inoculum-based forecasts are best suited to situations where the frequency of a problem is sporadic or spatially patchy, since diseases occurring every year will become routinely treated by farmers and rare diseases will not justify sampling expense. In addition, such a system is only necessary if there is no accurate weather-based forecasting system, if there is the possibility of severe economic yield loss and if there is a relatively cheap, rapid and reliable diagnostic method available. Finally the spatial variability of the target spores in relation to disease occurrence should be researched to optimise sampler location.

Optimal deployment of air samplers to assist disease control decisions is likely to vary according to the pathogen and crop system. It is not usually possible to use most types of air samplers for biosecurity purposes to detect very rare influx of an exotic species from a distant source due to dilution in the atmosphere (Jackson and Baylis 2011). However, some high volume spore traps such as the Jet spore trap, which samples 600 L/min (Burkard Manufacturing Co., Rickmansworth, UK) and the ionic spore trap (around 600 L/min) may be of use. Generally, once a pathogen has established a local sporulating focus of disease, it is more likely that the inoculum can be detected. The optimal deployment of air samplers varies according to the volume of air sampled by the device used, how widespread or common the pathogen is and the importance of the crop. Spore concentrations decline with distance from the source, usually a negative exponential or power function but other functions have been described. Usually these are similar in shape close to the source and vary from site to site and daily, according to wind speed, turbulence and crop canopy density. Usually it is not known exactly where the spore source is, and therefore which part of the concentration decline curve the sample was taken at. A relatively high concentration of spores could be caused by a very large distant source or a small source of spores very close to the sampler. As a result, care should be taken to interpret thresholds of spore concentrations to trigger disease control operations and normally this cannot be based on results from a single air sampler. However, some buffering against sampling effects, caused by releases of spores close to the sampler, can be achieved by mounting air samplers well above the ground or even on the roof of a tall building. For common plant pathogens it is possible to infer presence of airborne inoculum over a regional scale from a single air sampler located at rooftop height.

As farms become larger and growers become more reliant on mechanised equipment, the need for automated and user friendly diagnosis tools will be greater. Grower-friendly methods of pathogen detection need to be practical, readily available and cost effective. Methods to detect airborne spores of plant pathogens are becoming increasingly feasible due to advances in DNA-based diagnostics, antibody-based diagnostics, biosensors, and wireless communications. Diagnostic methods include spectral, fluorescence and image analysis methods for detection and identification of spores in air (or water) at the microscopic scale, but also DNA-based methods (PCR, qPCR, isothermal DNA amplification, and next generation sequencing particularly for bacteria, viruses and viroids), antibody-based methods (fluorescence microscopy and resonance imaging, ELISA, lateral flow devices, and biosensors such as holographic or SRi sensors) and biomarker-based methods (such as detection of volatile or particulate toxins or other metabolites by electrochemical biosensor). These are discussed below.

Culturing and Microscopy

A classical method for identification of fungal and oomycete spores is to use diagnostic keys with microscopy. In many cases it is only possible to identify spores to a genus and additional methods such as inducing spore germination to observe the germ tube branching pattern may be needed by sampling onto an impaction surface coated with a thin film of media. Spores can also be taken from culture plates used with semi selective media in impactors that collect spores onto agar plates, such as the Andersen sampler (West et al. 2008) or the Burkard portable air sampler (http://www.burkard.co.uk/portsamp.htm). Typically, a sample is placed into a humid chamber with or without a light source such as near-UV light depending on the species to encourage sporulation. However, these methods can be very time consuming, and impossible for obligate pathogens. Various staining methods, including immuno-labelling can be used to aid identification of spores. Visual recognition systems are available for automatic identification and counting of microscopic particles, particularly plant spores and pollen (www.aeromedi.org/home) and are being developed for common fungal spores. However, it remains extremely difficult to identify many plant pathogens to the species level. Unless a bioassay is used (e.g. inoculation onto a differential set of host plants), it is not possible to determine qualities that may be needed for quarantine purposes such as mating type, biotype or virulence group, for which nucleic acid-based methods are needed (described below).

Fluorescence and Particle Recognition Systems

In addition to methods used in combination with microscopy, systems used for automatic identification and counting of spores can be based on combinations of their optical properties such as size, shape, light scatter, pigmentation and UV fluorescence. These spore qualities can be assessed by passing airborne spores between a laser or LED source and appropriate sensors (Stanley et al. 2011) or spores can also be captured into liquid and processed into flow cytometry equipment for similar optical assessment (Day et al. 2002). Other devices that are mainly used by the military can detect microbes in samples (typically air) in near real-time. Most are not able to identify any specific organism but simply detect an elevated level of any viable microbes in the air. For example the Biotrace BBDS system (http://www.adpsa.co.za/Biotrace/Biotrace%20Intro.htm#Biotrace) uses microbial

ATP in the cells or spores to produce fluorescence from the luciferase enzyme derived from fireflies. Other systems can detect a single organism such as *Bacillus anthcis* (Anthrax spore sensor; http://www.nano.org.uk/news/733/). These approaches offer real time detection of biological particles but usually are not specific and at best will indicate only a type of spore rather than a species. They are better applied to clean-room, clinical and bioterrorism applications where real-time detection of a potential threat is necessary but could play a role in targeting use of more specific diagnostic methods.

Nucleic Acid-Based Diagnostics

Typically an air sample may need cell lysis (e.g. by shanking with microscopic glass beads) and DNA extraction and purification to make DNA available for an assay, but some species of fungi and bacteria, particularly with delicate spores, can be detected directly by PCR (Williams et al. 2001). Various nucleic acid based assays have been developed, generally using a probe or primers that bind to a specific sequence of DNA or RNA, present in the target. This sequence will be specific to a taxon such as a kingdom, species, or a genetic trait present in a population. An amplification step may or may not be used to enhance the detection of the target sequence. The main methods involving amplification steps are polymerase chain reaction or PCR and quantitative- or real-time PCR, which have been used extensively since the late 1990s for identification and quantification of plant pathogens and the study of genetic traits such as fungicide resistance or toxin production. PCR requires the design and testing for specificity and sensitivity of primers, either by sequencing DNA of one or more isolates or searching sequence databases. PCR methods use Taq polymerase and a cocktail of nucleotides plus potassium and magnesium or manganese ions in a reaction mixture with the sample DNA. The process involves thermal cycling, typically heating to 90 °C to separate all double-stranded DNA, followed by rapid cooling to more moderate temperatures (usually 50–70 °C) to allow binding of primers (forward and reverse) to specific sections of DNA that complement their designed sequence, only if the target sequence is present. This is followed by subsequent extension of the bound primer mediated by Taq polymerase to make a double stranded product. Multiple cycles of the specific thermal regime, creates a mass of PCR product or amplicon, which can be visualised on a gel by electrophoresis. For qPCR, addition of a DNA-specific fluorescent dye such as SYBR green allows quantification of the amplicon by fluorescence measurement after each replication cycle and this is compared to the fluorescence of known amounts of the pure pathogen's DNA which are run at the same time as standards. A better qPCR method for improved specificity is the Taqman qPCR, which uses in addition to primers, an oligonucleotide probe which is labelled at the 5′ end with a fluorophore and has a 'quencher' molecule on the 3′ end that prevents fluorescence while the quencher is in close proximity to the fluorophore. The probe binds to the amplicon but is cleaved into separate nucleotides by the polymerase that is extending from the 5′ to 3′ direction, which separates the fluorophore and quencher to allow fluorescence, which is measured after each cycle (e.g. Yang et al. 2004). Alternatively nucleic acids and particularly PCR products can be visualised by other fluorescent probes such as scorpions (Sharkey et al. 2004) and molecular beacons (Tyagi and Kramer 1996). These can be multiplexed so that different Taqman, scorpion or molecular beacon probes with different fluorescent labels can be used to quantify different nucleic acid targets in one test (Sharkey et al. 2004) allowing a single sample to be tested for the presence of several pathogens. With these techniques, because PCR can be inhibited by the presence of certain chemicals or the DNA extraction process may not have worked, care is needed to include controls (Peccia and Hernandez 2006; McDevitt et al. 2007). This can comprise testing a sub-sample of DNA with consensus fungal primers that will detect all fungi, which are to be expected in any outdoor air sample, or spiking a sub-sample with a known amount of the target DNA to test for inhibition.

The use of isothermal DNA amplification methods such as Loop-mediated isothermal amplification (LAMP; Notomi et al. 2000) and recombinase polymerase amplification (Piepenburg et al. 2006; www.TwistDX.com) is increasingly becoming a tool for pathogen detection. These methods use enzymes to separate double stranded DNA without the need for heating the sample to 90 ºC as with traditional PCR and therefore are more analogous to DNA replication in living cells. This allows the reaction to take place at a single temperature, typically around 65 °C for LAMP and 37 °C for the TwistDX method. This makes the method more suitable for lightweight, portable devices as heating small tubes to a set temperature is relatively easy compared to thermal cycling and has lower energy requirements. A hand-held device is currently under development (Li et al. 2013). Although the technique may not be quite as sensitive as qPCR, it is often possible to be used on relatively crude DNA extracts, which also facilitates on-site detection using portable equipment. A deposit of air particulates, for example, can be lysed chemically in an extraction buffer and then the liquid can be used in the reaction (Boonham et al. 2013; Li 2013; Mahaffee et al. 2013). Results are obtained typically in 5–10 min. PCR and LAMP can be used to detect RNA (for detection of viruses or active gene expression in eukaryotes) by using a reverse-transcriptase to double the single-stranded RNA into DNA (known as RT-PCR; Freeman et al. 1999).

DNA and Protein Arrays and Microarrays

DNA microarrays consist of a solid surface, such as a glass slide, onto which is printed a known arrangement of tiny dots of nucleic acid primers. They are commonly known as DNA-chips or DNA-arrays. The dots of primers each bind to specific nucleic acids from an air or other environmental sample, or to different genes of a target species under investigation. As a result, DNA of an environmental sample containing multiple species and genotypes of particular species can be tested in one assay. DNA multiscan is one example, which is able to detect multiple plant pathogenic bacteria, oomycetes and fungi (http://www.dnamultiscan.com/en/home.html). Similarly, species specific protein targets can also bind to arrays of different antibodies, bound to a solid surface in a similar way to that described for ELISA (below). The term 'lab on-a-chip' has been used for systems that miniaturise processes used in diagnostics such as DNA or protein arrays. These are microfluidic devices in which liquids are moved through tiny pathways carved into glass slides. It is claimed that this gives faster reactions due to tiny volumes used taking advantage of diffusion and large surface area to volume ratios and ability to heat sections of the chip using minimal power.

Genomics-Based Detection

This approach is certainly not a rapid test but is of particular interest for identification of unknown bacteria, viruses and viroids, especially because no previous sequence data of the organism is needed, nor specific primers or probes and there is also no need to culture the organism (Rodoni et al. 2013), which is important since only around 10 % of bacteria are culturable (Pace 1997). Next generation sequencing (NGS) (various platforms exist such as Solexa, 454 Roche, Illumina and Ion Torrent) pyrosequencing and metagenomics have been used (Rodoni et al. 2013; Hopkins et al. 2013). Currently, sequencing to identify an unknown pathogen can cost as little as $850 and takes about 2 weeks (Olmos et al. 2013).

Immunology-Based/Antibody-Based Detection Methods

Advances in rapid and cost effective antibody production over the last 10 years have allowed the development of antibody based detection systems for plant disease diagnosis. Antibody based diagnosis/detection systems have been designed to be used in the laboratory as well as in hand held devices to be used for on-site detection. These systems are based on the use of antibodies as high affinity ligands which will bind to species specific cell surface fragments or antigens or even whole cell substrates. Antibodies that can be used in various detection systems can be monoclonal or polyclonal. Monoclonal antibodies (mAb) will bind to a single site or epitope of the target fragment whereas polyclonal antibodies (pAb) will bind to multiple epitopes on a single antigen allowing more specificity for target detection. Routinely, mAbs and pAbs are produced by the injection of the whole cell/pathogen or surface fragment into a suitable animal. An increasingly popular antibody production method is the use of bacterial expressed recombinant fragments which include single chain variable fragments (scFv) (Skottrup et al. 2008). These fragments are a sixth of the size of standard antibodies (Lamberski et al. 2006) and maintain high specificity to the parental mAb. Using bacterial cultures for large

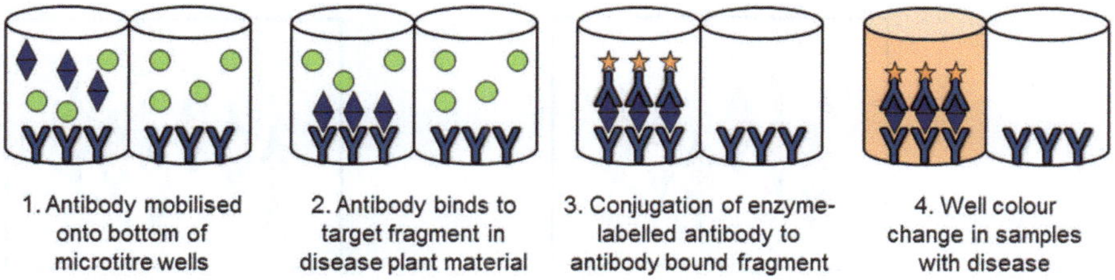

Fig. 1 Stages in an ELISA assay

scale antibody production ensures an infinite supply of genetically stable scFvs, which is cost effective and more attractive than using animal systems. Antibody based kits for plant disease diagnosis were originally developed in 1977 (Clark and Adams 1977) but are now commercially available for laboratory use. Most companies supply a DAS-ELISA (double antibody sandwich enzyme-linked immunosorbent assay) kit for disease diagnosis. This test involves the use of multi well microtiter plates which have a specific antibody adsorbed to surface of microtiter wells. A sample is incubated for a period of time in an extraction buffer in the well. Incubation allows the binding of the target antigen to the pathogen specific antibody. A conjugate buffer containing an enzyme-labelled antibody is then added to the wells and incubated. After a final washing step, another buffer reacts with the conjugated enzyme labelled antibody, which is bound to the target fragment. Those wells that contained samples with the pathogen of interest will change colour and can be visualised using a spectrophotometer (Fig. 1). There are many commercial ELISA kits available for the detection of plant viruses because specific antibodies can be raised easily against such organisms. Fewer kits are available for fungal pathogen diagnosis to a species level. This is partially because of cross reactivity between different species and in the case of most spores, due to low allergenicity of the surface coat, which may be overcome by incubating the sample to induce spore germination to target antigens on- or secreted by mycelium. Some ELISA kits have however been developed for fungi and oomycetes including, *Botrytis cinerea*, *Pythium* spp., *Phytophthora spp* and *Septoria spp (Agdia, Bioreba, Neogen).*

ELISA kits allow for quantitative and accurate diagnosis of the disease, however the tests require skilled technicians and specialised equipment. There is also usually a time delay for the grower to obtain results which may limit the time window for successful disease control. ELISA technology, although not thought of as traditional biosensor technology, can be classified as a biosensing technique because it transforms the biological response of the antibody binding event and relays the response optically with the use of a spectrophotometer to detect changes in chemilluminescence or fluorescence. The micro titre immune spore trap (MTIST; Burkard Manufacturing Co., Rickmansworth, UK) has been designed specifically, as the name suggests, to sample airborne particles such as spores directly into microtiter plates to facilitate ELISA (Kennedy et al. 2000).

Handheld lateral flow devices (LFD) have been developed so that growers can carry out rapid disease assessments in the field as opposed to waiting for results from a diagnostics laboratory. One step LFDs are similar to ELISA assays as they

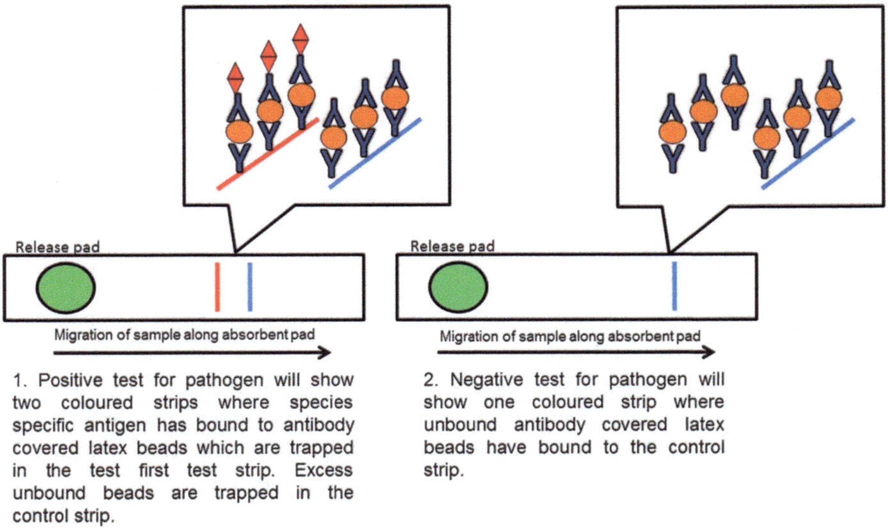

Fig. 2 Positive and negative results of a lateral flow device. Migration of sample along absorbent pad

use pAbs and mAbs to bind to a pathogen specific fragment or antigen, however they are not fully quantitative. A LFD commercially available from Pocket Diagnostics^{TM} (www.pocketdiagnostic.com) consists of antibody-coated latex beads which will bind the specific pathogen antigen absorbed from the plant extract (Ward et al. 2004). The test works by the grower/user taking a sample (such as an infected leaf) and crushing it up in a bag, but the same technique can be applied to air sample particulates, especially if incubated to allow germination. They then incubate the sample in a buffer. After incubation, the buffer solution is dropped onto a release pad on the bottom of the LFD which contains specific antibody-coated latex beads that will bind to the target antigen from the plant extract. The solution containing the conjugated antibodies will migrate along an absorbent pad to a test strip where latex beads containing the bound antigen will be trapped forming a visible line on the pad. Surplus unbound antibodies migrate further along the pad and are trapped onto a second strip (Fig. 2). This acts as a control indicating that the test worked correctly. There are a variety of LFDs that are capable of identifying a various plant pathogens including *B. cinerea*, oomycetes such as *Pythium* and *Phytophthora spp* and bacterial pathogens including *Ralstonia solanacearum* and *Erwinia amylovora*.

Antibodies for Metabolite Sensing

Plant pathogen detection devices can also target secreted metabolites rather than cell surface proteins for accurate diagnosis. A rapidly expanding example of this is the need for detection methods for fungal and bacterial toxins that are secreted into substrates and may become airborne in dusts, particularly in grain stores and processing areas. Many species of fungi including *Penicillium spp* and *Aspergillus spp* secrete mycotoxins into the food substrate they are growing on. Some of these toxins, if eaten, can have harmful effects including sickness and diarrhoea, gastro-intestinal problems, and kidney damage or in some cases are carcinogenic. Manufactures and sellers of perishable foods have to follow strict legislation to ensure their food does not contain mycotoxin levels above a certain threshold outlined by the EU Commission Regulation (EC) No. 1881/2006. On site diagnosis kits for many toxins including aflatoxins, ochratoxins and DON, have been developed for LFDs, however lab based ELISA and HPLC testing is also used to quantify amounts of the toxin more accurately.

Biosensors for Pathogen Detection

Since the invention and success of the blood glucose biosensor (Clark and Lyons 1962), research into using biosensing as a way to monitor and quantify elements of biological systems at point of care (POC) has flourished in medical professions. Biosensing applications to detect plant pathogens in an agricultural system is an expanding field as there is the potential to develop devices to detect pathogens rapidly and in an automated format that can be easily used and will require little technical skill. There is also the potential to combine biosensor outputs with wireless networks to enable data indicating detection events to be sent to a central monitoring and processing unit, which can then incorporate the information into a disease forecasting system. However, much of the technology is lab based and remains to be translated into handheld or automated devices.

A biosensor takes a biological response and translates it into an electrical signal (Turner 2000). It is an analytical device, which integrates a biological sensing element or bioreceptor within a physicochemical transducer and is required to produce an electronic signal proportional to the specific analyte that it is measuring (Nierman et al. 2005). Bioreceptors include antibodies, enzymes, whole cells and DNA (Fig. 3). These are usually immobilised onto some form of sensor surface.

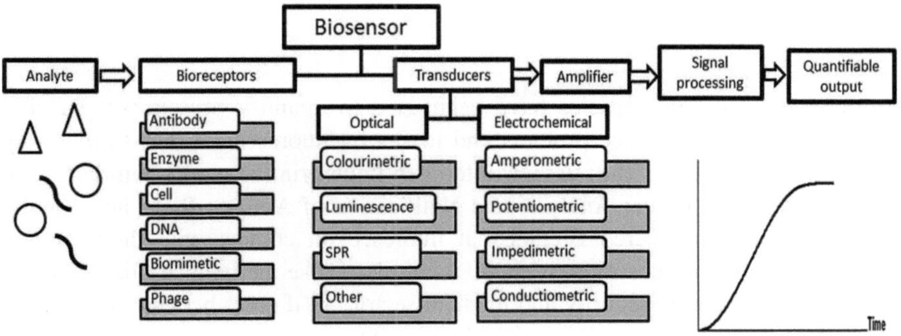

Fig. 3 Stages in the operation and output of a wide range of possible biosensors

The transducer, which can be optical or electrochemical, takes the biological response (such as a change of light-pathway or electrical conductance caused by binding of an antigen to an antibody, or an enzyme mediated chemical reaction), and changes it into an electrical signal. An example of optical transducers includes surface plasmon resonance (SPR) and holographic biosensors or fibre optics. Electrochemical transducers make use of changes in current, potential, impedance and conductance across an electrode surface for detection events (Velusamy et al. 2010).

Surface Plasmon Resonance (SPR)

A popular technique to characterise the interactions of small molecules such as proteins, polysaccharides and nucleic acids is SPR. It is a label free, optical biosensing technique that can be used to detect molecular binding events. SPR incorporates the use of a light source, which passes light through a prism and hits a gold surface sensor chip. The light bounces off the sensor chip and hits a detector. At a certain angle known as resonance angle, light is absorbed by the sensor film, causing electrons or surface plasmons to resonate (Skottrup et al. 2008). This causes a loss of intensity in the reflected beam and can be detected by a reflectivity curve (Biosensing Instrument Inc ©). Antibodies can be immobilised onto sensor chip surfaces and the binding events at the sensor surface causes a change in the refractive index, which is then monitored. This is a real-time method of detection and can also be used to calculate rates of binding events. SPR is still very much a lab based method of detection but has been used to detect a variety of viruses including Cowpea mosaic virus (Torrance et al. 2006) and Lettuce mosaic virus (Candresse et al. 2007). Fungal pathogens including *Fusarium culmorum* (Zezza et al. 2006) and *Aspergillus niger* (Nugaeva et al. 2007) and the oomycete *Phytophthora infestans* (Skottrup et al. 2007) have been detected by SPR in a laboratory. Many of these detection systems use a Biacore SPR sensor surface onto which species specific antibodies are immobilised. Although SPR technology has its advantages as a real time, label free detection method, the main disadvantage is that it lacks sensitivity for measuring small molecules.

Other Antibody-Based Biosensors

Antibody-coated nanoparticles have been used to quantify presence of specific antigens, including surface proteins and toxins by fluorescence in a similar way to an *in-situ* ELISA (http://www.intelligentfingerprinting.com/) similarly the canary® sensor (Cellular Analysis and Notification of Antigen Risks and Yields) are genetically engineered B-cells that fluoresce on contact with their antigen (http://innovativebiosensors.com/about_ibi.html). These methods are being developed for use in the biosecurity and medical sector but if costs become reasonable, have great potential to be applied for plant health.

Electrochemical Biosensors

Electrochemical transduction is a popular choice for biosensor systems and has been researched thoroughly. This form of transduction was used in the original glucose biosensor system. These biosensors can be classified into amperometric, potentiometric, impedimetric and conductometric, based on the observed factors such as current, potential, impedance and conductance respectively (Velusamy et al. 2010). For example an amperometric biosensor consists of an oxidoreductase enzyme which is stabilised onto an electrode. The enzyme is specific to a particular analyte, which is specific to the pathogen being detected. Upon arrival of the analyte, the enzyme will oxidise the analyte and the electrons generated in the reaction are shuttled to the electrode through artificial electron acceptors or mediators such as ferrocene or hexacyanoferrate. The mediation of electrons produces a current that is directly proportional to the concentration of the analyte (Turner 2000). Antibodies can also be stabilised onto a biosensor so that the binding event of target antigen to antibody can be detected. Electrochemical biosensors have been successfully used to diagnose many food borne pathogens including *E-coli* and *Salmonella*, which can cause significant harm to humans and animals if ingested (Yang et al. 2001; Muhammad-Tahir and Alocilja 2003, 2004). One study used an amperometric biosensor to detect *Salmonella typhimurium* cells. The bacterial cells were bound by magnetic- beads, which were conjugated to antibodies and then subsequently detected by an alkaline-phosphatase (AP)-labelled anti-*Salmonella* antibody. The AP antibody catalysed the breakdown of *para*-aminophenyl phosphate into electro-active *para*-aminophenol. This generates an electrochemical signal and can measure over 8×10^3 bacterial cells/mL (Gehring et al. 1996).

Networked Air-Samples in Crop Protection

SYield is a system of air-sampler nodes designed to work with computer models to provide a sensor network for pathogen detection and prediction. The first generation of the SYield platform (Fig. 4) warns growers of imminent risk of *Sclerotinia* in oilseed rape and works when airborne spores of *Sclerotinia* are sampled into a liquid growth medium and after a short period of incubation, secrete oxalic acid (a pathogenicity factor), which is detected by an electro-chemical biosensor. Weather data are collected by the same sample devices and data transmitted by mobile phone signal for integration into a risk alert (http://www.syield.net/home.html). The air sampler used in this device is a miniature virtual impactor, which allows collection of particles into liquid for sample periods up to 24 h with minimal evaporation of the collection liquid.

Fig. 4 SYield automated spore trap under validation with conventional air samplers at Rothamsted Research

Conclusions and Future Prospects

Many diagnostics used with air sampling in biosecurity and crop protection are under rapid development. There is an increasing need for in-field diagnostics that can use the above technologies to detect pathogens automatically, without the need for specialised equipment or technical experts. Crossover technologies from the biomedical and biosecurity sectors are allowing the development of automated on-site detection kits, which aid growers in rapid plant pathogen identification so they can implement timely integrated pest control strategies. With advances in sequencing technologies, the number of species specific nucleic acid-based assays and even protein-based assays will allow current methods of detection to be expanded for identification of more pathogen species. There are various trade-offs concerning cost, accuracy, specificity and ease of use in the methods outlined above. Clearly these methods will be augmented by new developments and are likely to become adopted by researchers and industry as costs reduce in a similar way to the uptake of home computers or mobile phones. Stringency and robustness of specificity and accuracy are key attributes in the development of any new diagnostic and detection technology and so thorough testing on a wide range of isolates is essential. Ideally new devices should be adaptable to detect different organisms using the same air sampling hardware, simply by changing consumables such as primers, probes, antibodies or biosensor. Ultimately, the use of advanced but cheap diagnostic methods will extend to automated sampling devices with wireless reporting. Automated sampling devices of the future will also assist plant health inspectors in monitoring the arrival of new pathogens that may pose a threat to both agricultural systems and natural environments.

Acknowledgments The authors thank the following funders: Technology Strategy Board (UK), Syngenta and the SYield consortium (http://www.syield.net/home.html), the BBSRC (UK), the HGCA, Defra and the European Union Seventh Framework Programme (FP7/2007–2013) under the grant agreement 265865 (PURE project).

References

Boonham N (2013) Q-DETECT – developing pest and pathogen detection methods for national plant protection organizations. Acta Phytopathol Sinica 43(Suppl): 278 [International Congress of Plant Pathology 2013 abstract] O22.003

Brown JKM, Hovmøller MS (2002) Aerial dispersal of pathogens on the global and continental scales and its impact on plant disease. Science 297:537–541

Candresse T, Lot H, German-Retana S, Krause-Sakate R, Thomas J, Souche S, Delaunay T, Lanneau M, Le Gall O (2007) Analysis of the serological variability of Lettuce mosaic virus using monoclonal antibodies and surface plasmon resonance technology. J Gen Virol 88:2605–2610

Clark MF, Adams AN (1977) Characteristics of microplate method of enzyme-linked immunosorbent assay for the detection of plant-viruses. J Gen Virol 34:475–483

Clark LC, Lyons C (1962) Electrode systems for continuous monitoring in cardiovascular surgery. Ann N Y Acad Sci 102(1):29

Cox CS, Wathes CM (eds) (1995) Bioaerosols handbook. CRC Press, Boca Raton, pp 623

Day JP, Kell DB, Griffith GW (2002) Differentiation of Phytophthora infestans sporangia from other airborne biological particles by flow cytometry. Appl Environ Microbiol 68:37–45

Fraaije BA, Cools HJ, Fountaine J, Lovell DJ, Motteram J, West JS, Lucas JA (2005) QoI resistant isolates of *Mycosphaerella graminicola* and the role of ascospores in further spread of resistant alleles in field populations. Phytopathology 95:933–941

Freeman WM, Walker SJ, Vrana KE (1999) Quantitative RT-PCR: pitfalls and potential. Biotechniques 26:112–122

Fröhlich-Nowoisky J, Pickersgill DA, Després VR, Pöschl U (2009) High diversity of fungi in air particulate matter. Proc Natl Acad Sci U S A 106:12814–12819

Gehring AG, Crawford CG, Mazenko RS, VanHouten LJ, Brewster JD (1996) Enzyme-linked immunomagnetic electrochemical detection of Salmonella typhimurium. J Immunol Methods 195:15–25

Gottwald TR, Graham JH, Egel DS (1992) Analysis of foci of Asiatic citrus canker in a Florida citrus orchard. Plant Dis 76:389–396

Gottwald TR, Hughes G, Graham JH, Sun X, Riley T (2001) The citrus canker epidemic in Florida: the scientific basis of regulatory eradication policy for an invasive species. Phytopathology 91:30–34

Gregory PH (1952) Spore content of the atmosphere near the ground. Nature 170:475

Gregory PH (1973) The microbiology of the atmosphere, 2nd edn. Leonard Hill, Aylesbury, pp 377

Hopkins AJM, Castanyo C, Boberg JB, Stenlid J (2013) Methods for the early detection of new invasive forest pathogens. Acta Phytopathol Sinica 43(Suppl):85 [International Congress of Plant Pathology 2013 abstract] O04.016

Jackson SL, Bayliss KL (2011) Spore traps need improvement to fulfil plant biosecurity requirements. Plant Pathol 60:801–810

Kennedy R, Wakeham AJ, Byrne KG, Meyer UM, Dewey FM (2000) A new method to monitor airborne inoculum of the fungal plant pathogens Mycosphaerella brassicicola and Botrytis cinerea. Appl Environ Microbiol 66:2996–3003

Lacey ME, West JS (2006) The Air Spora. Springer, Dordrecht, The Netherlands, pp 156

Lamberski JA, Thompson NE, Burgess RR (2006) Expression and purification of a single-chain variable fragment antibody derived from a polyol-responsive monoclonal antibody. Protein Expr Purif 47(1):82–92

Li X, Nie J, Arsenault H, Xu H, De Boer SH (2013) On-the-spot diagnosis and detection of plant pathogens. Acta Phytopathol Sinica 43(Suppl): 84 [International Congress of Plant Pathology 2013 abstract] O04.014

Mahaffee W, Keune J, Thiessen L, Grove G (2013) Bringing nucleic acid based detection of pathogens to the masses: the use of isothermic amplification approaches by practitioners. Acta Phytopathol Sinica 43(Suppl): 278 [International Congress of Plant Pathology 2013 abstract] O22.002

McCartney HA, Fitt BDL, West JS (2006) Dispersal of foliar fungal plant pathogens: mechanisms, gradients and spatial patterns. In: Cooke BM, Jones DG, Kaye B (eds) The epidemiology of plant diseases, 2nd edn. Springer, Dordrecht, pp 159–192

McDevitt JJ, Lees PSJ, Merz WG, Schwab KJ (2007) Inhibition of quantitative PCR analysis of fungal conidia associated with indoor air particulate matter. Aerobiologia 23:35–45

Morris CE, Monteil CL, Berge O (2013) Novel understanding of the water cycle as a link between unsuspected habitats of airborne pathogens – what consequences for plant disease management? Acta Phytopathol Sinica 43(Suppl):17 [International Congress of Plant Pathology 2013 abstract] O01.004

Muhammad-Tahir Z, Alocilja EC (2003) Fabrication of a disposable biosensor for Escherichia coli O157: H7 detection. IEEE Sens J 3(4):345–351

Muhammad-Tahir Z, Alocilja EC (2004) A disposable biosensor for pathogen detection in fresh produce samples. Biosyst Eng 88(2):145–151

Nierman WC, Pain A et al (2005) Genomic sequence of the pathogenic and allergenic filamentous fungus Aspergillus fumigatus. Nature 438(7071):1151–1156

Notomi T, Okayama H, Masubuchi H, Yonekawa T, Watanabe K, Amino N, Hase T (2000) Loop-mediated isothermal amplification of DNA. Nucleic Acids Res 28:e63

Nugaeva N, Gfeller KY et al (2007) An antibody-sensitized microfabricated cantilever for the growth detection of Aspergillus niger spores. Microsc Microanal 13(1):13–17

Olmos A, Bertolini E, Martínez MC, Candresse T, Glasa M, Pina JA, Cambra M (2013) Deep sequencing: a powerful tool for detection and characterization of known and new plant viruses and viroids. Acta Phytopathol Sinica 43(Suppl): 280 [International Congress of Plant Pathology 2013 abstract] O22.010

Pace NR (1997) A molecular view of microbial diversity and the biosphere. Science 276:734–740

Pady SM, Kapica L (1955) Fungi in air over the Atlantic Ocean. Mycologia 47:34–50

Peccia J, Hernandez M (2006) Incorporating polymerase chain reaction-based identification, population characterization, and quantification of microorganisms into aerosol science: a review. Atmos Environ 40:3941–3961

Piepenburg O, Williams CH, Stemple DL, Armes NA (2006) DNA detection using recombination proteins. PLoS Biol 4:1115–1121

Rodoni B, Mann R, Smith G, Chapman T, Bellgard M, Stack J (2013) Genomics-based detection of bacterial plant pathogens; Implications for quarantine and trade policy. Acta Phytopathol Sinica 43(Suppl):82 [International Congress of Plant Pathology 2013 abstract] O04.008

Sharkey FH, Banat IM, Marchant R (2004) Detection and quantification of gene expression in environmental bacteriology. Appl Environ Microbiol 70(7):3795–3806

Skottrup P, Nicolaisen M, Justesen AF (2007) Rapid determination of Phytophthora infestans sporangia using a surface plasmon resonance immunosensor. J Microbiol Methods 68(3):507–515

Skottrup PD, Nicolaisen M, Justesen AF (2008) Towards on-site pathogen detection using antibody-based sensors. Biosens Bioelectron 24(3):339–348

Stanley WR, Kaye PH, Foot VE, Barrington SJ, Gallagher M, Gabey A (2011) Continuous bioaerosol monitoring in a tropical environment using a UV fluorescence particle spectrometer. Atmos Sci Lett 12:195–199

Torrance L, Ziegler A, Pittman H, Paterson M, Toth R, Eggleston I (2006) Oriented immobilisation of engineered single-chain antibodies to develop biosensors for virus detection. J Virol Methods 134(1–2):164–170

Turner APF (2000) Biochemistry – biosensors sense and sensitivity. Science 290 (5495):1315–1317

Tyagi S, Kramer FR (1996) Molecular beacons: probes that fluoresce upon hybridization. Nat Biotechnol 14(3):303–308

Van de Wouw AP, Stonard JF, Howlett B, West JS, Fitt BDL, Atkins SD (2010) Determining frequencies of avirulent alleles in airborne Leptosphaeria maculans inoculum using quantitative PCR. Plant Pathol 59:809–818

Velusamy V, Arshak K, Korostynska O, Oliwa K, Adley C (2010) An overview of foodborne pathogen detection: in the perspective of biosensors. Biotechnol Adv 28(2):232–254

Ward E, Foster SJ, Fraaije BA, McCartney HA (2004) Plant pathogen diagnostics: immunological and nucleic acid-based approaches. Ann Appl Biol 145(1):1–16

West JS, Atkins SD, Emberlin J, Fitt BDL (2008) PCR to predict risk of airborne disease. Trends Microbiol 16:380–387

Williams RH, Ward E, McCartney HA (2001) Methods for integrated air sampling and DNA analysis for detection of airborne fungal spores. Appl Environ Microbiol 67:2453–2459

Yang LJ, Ruan CM, Li YB (2001) Rapid detection of Salmonella typhimurium in food samples using a bienzyme electrochemical biosensor with flow injection. J Rapid Methods Automat Microbiol 9(4):229–240

Yang DK, Kweon CH, Kim BH, Lim S-I, Kim S-H, Kwon J-H, Han H-R (2004) TaqMan reverse transcription polymerase chain reaction for the detection of Japanese encephalitis virus. J Vet Sci 5(4):345–351

Zezza F, Pascale M, Mule G, Visconti A (2006) Detection of Fusarium culmorum in wheat by a surface plasmon resonance-based DNA sensor. J Microbiol Methods 66(3):529–537

siRNA Deep Sequencing and Assembly: Piecing Together Viral Infections

Jan Kreuze

Abstract RNA silencing constitutes a fundamental antiviral defense mechanism in plants in which host enzymes cut viral RNA into pieces of 20–24 nt. When isolated, sequenced en-mass and properly assembled or aligned these virus-derived small RNA (sRNA) sequences can reconstitute genomic sequence information of the viruses being targeted in the plant. This approach is independent of the ability to culture or purify the virus and does not require any specific amplification or enrichment of viral nucleic acids as it automatically enriches for small RNAs of viral origin by tapping into a natural antiviral defense mechanism. Using this technique known and novel DNA and RNA viruses as well as viroids have been identified at sensitivity levels comparable to PCR. This chapter will examine the strength and caveats of small RNA sequencing and assembly (sRSA), utilizing examples from literature as well as our own unpublished experiences and analysis of publically available plant sRNA sequence datasets.

Keywords Diagnostics • Small-RNA • Sequencing • Assembly • Virus • Viroid • Detection • Plants

Introduction

Next generation sequencing (NGS) technologies have led to a revolution in sequencing based applications in biological sciences. Viral diagnostics has not lagged behind with a number of NGS based approaches for pathogen identification emerging over the last 5 years. In principal NGS could enable universal diagnostics, allowing the identification of any pathogen in a sample based on its sequence. However even though throughput is ever increasing, sensitivity can be an issue

J. Kreuze (✉)
Laboratory of Virology, Peru International Potato Center (CIP),
Avenida la Molina, 1895 La Molina, Peru
e-mail: j.kreuze@cgiar.org

M.L. Gullino and P.J.M. Bonants (eds.), *Detection and Diagnostics of Plant Pathogens*,
Plant Pathology in the 21st Century 5, DOI 10.1007/978-94-017-9020-8_2,

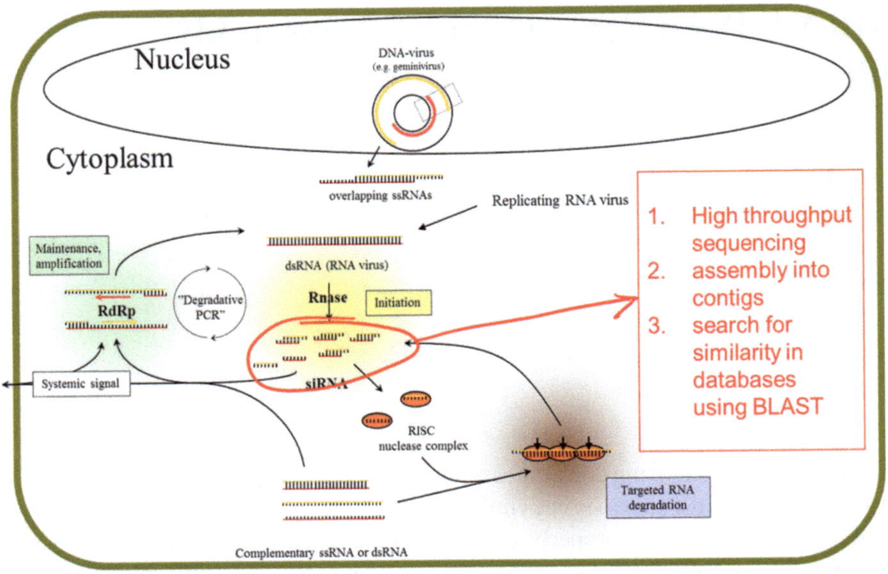

Fig. 1 Antiviral RNA silencing leads to the accumulation of virus derived siRNAs, which can be isolated, sequenced and assembled to posteriorly identify viruses by homology to known virus sequences

when low titer pathogens are involved, especially when taking price into consideration. While different sequencing platforms have been used, such as 454 and Illumina, which may influence results due to differences in sequence length and volume, key differences lie in type of nucleic acids that have been used for sequencing as well as methods to enrich for viral nucleic acids. The most straight forward approach is probably simply to sequence total mRNA and this approach has been successfully applied to identify a number of viruses (Wylie and Jones 2011; Wylie et al. 2012a, b, 2013; Al Rwahnih et al. 2009). Adams et al. (2009) used RNA from a healthy plant for subtractive hybridization, thereby enriching the RNA for pathogen sequences. Other studies sequenced purified double stranded RNA (dsRNA) (Al Rwahnih et al. 2011; Coetzee et al. 2010; Roossinck et al. 2010; Candresse et al. 2013) and a combination of both techniques has also been applied (Adams et al. 2013). NGS has also been combined with rolling circle amplification to efficiently identify circular DNA plant viruses (Hagen et al. 2012; Ng et al. 2011).

The most extensively utilized approach to date however relies on enrichment of viral sequences by the plant itself (Kreuze et al. 2009): small RNA sequencing and assembly (sRSA; Fig. 1). The approach exploits a natural anti-viral defence system called RNA silencing or RNA interference (RNAi), which is present in all eukaryotic organisms and generates 21–24 nt small RNA (sRNA) molecules homologous to viruses that may be infecting them (Mlotshwa et al. 2008). An additional benefit of this approach is that it makes ideal use of sequencing platforms that generate enormous quantities, but short sequences. It has been successfully employed to identify many new plant viruses of all possible genome types (Kashif et al. 2012; Loconsole et al. 2012a, b; Bi et al. 2012; Wu et al. 2012; Zhang et al. 2011; Li et al. 2012; Hwang et al. 2013; Cuellar et al. 2011; De Souza et al. 2013; Fuentes et al. 2012; Kreuze et al. 2009, 2013; Untiveros et al. 2010; Sela et al. 2012; Roy et al. 2013), including viroids (Li et al. 2012; Wu et al. 2012) but has also been applied to identify viruses from worms, insects and even humans (Huang et al. 2013; Ma et al. 2011; Isakov et al. 2011; Wu et al. 2010).

sRSA: An Historic Overview

Viruses are both inducers and targets of RNA silencing, a fundamental antiviral defense mechanism in eukaryotic organisms (Mlotshwa et al. 2008; Haasnoot et al. 2007) (Fig. 1). RNA silencing is a cytoplasmic cell surveillance system that recognizes double-stranded RNA (dsRNA) and specifically eliminates RNAs homologous to the inducer RNA by cleavage using RNase III endonucleases called dicers. Plants encode several Dicer-like enzymes that recognize and cleave long dsRNA molecules to 21-, 22-, and 24-bp fragments that act as small interfering RNAs (siRNAs). siRNAs bind to ribonuclease H–like proteins in the RNA induced silencing complex (RISC) and are used to detect homologous single-stranded RNA (ssRNA) molecules for cleavage, producing more siRNAs. In plants, RNAi becomes amplified when the cleaved RNA recruits an RNA-directed RNA polymerase to generate more dsRNA, which is again cleaved by a dicer protein to produce secondary siRNAs, that are once again able to detect and cleave homologous RNA in a type of 'degradative PCR' cycle (Fig. 1). This leads to the accumulation of large amounts of siRNAs with homology to the invading virus. Also DNA viruses have been found to induce the production of copious amounts of siRNAs in plants, although the exact mechanism by which the system is initially triggered is not yet fully elucidated. Because accumulation of virus-derived siRNAs appeared to be a common feature of defense against viral infection in diverse eukaryotic hosts, and seemed particularly abundant in plants, using these defense molecules, which could be isolated based on their size, to identify viruses infecting a host had become a plausible concept (Kreuze et al. 2009). Next generation sequencing platforms generating massive amounts, but rather short reads such as Illumina (Solexa) and SOLiD were rapidly developing, and with them a set of bioinformatics tools for *de novo* assembly of such short reads. However, even if these tools had been developed for short reads, they were generally longer than the 21–24 nucleotides of siRNAs, and it was not clear how well they would function using reads of that size. It was also unclear to what extent viral siRNAs overlap and thus whether they were suitable for the assembly into larger contigs that could be used for virus detection. Our study using experimentally infected sweetpotatoes (Kreuze et al. 2009) provided a positive answer to both questions since all three assembly tools tested (Velvet, VCAKE and SSAKE) were able to assemble viral contigs from five different viruses efficiently, even if a full genome could only be assembled *de novo* from one of the two RNA viruses (a potyvirus) and only after manual re-assembly. Although only relatively short contigs could be assembled for the other RNA virus (crinivirus), guide strand mediated assembly by alignment using MAQ and using the siRNA reads to the genome of a related virus isolate was able to assemble 98 % of the genome of that virus as well. A notable aspect of that study was that besides the two expected RNA viruses a number of additional contigs corresponding to DNA viruses (two strains of a new badnavirus and a mastrevirus) were identified and further confirmed by PCR and Sanger sequencing. Although we have subsequently been able to confirm that the badnaviruses are not genome-integrated and can be graft transmitted (Kreuze, unpublished), identification of sequences related to para-retroviruses may not always signify infection by a virus as they may represent silencing of endogenous integrated virus sequences (see section "Reverse transcribing viruses" of this chapter below).

Soon afterwards the principal was demonstrated in invertebrate animals by Wu et al. (2010), who were able to assemble a known virus and five previously described viruses from nematode, fruit fly and mosquitoes. Later Isakov et al. (2011) identified Human immunodeficiency virus (HIV) and a mycoplasma in human cells through the application of sRSA. Each of these studies identified unexpected novel pathogens underlining not only the ability to detect infectious agents without a priori knowledge, but also the fact that organisms considered to be healthy, may not always be virus free, including well studied cell lines.

Several studies have followed characterizing plant viruses with various types of genomes, including single- (Loconsole et al. 2012b; Kashif et al. 2012), and double- (Kashif et al. 2012; Zhang et al. 2011) stranded DNA genomes, positive- (Bi et al. 2012; Cuellar et al. 2011; De Souza et al. 2013; Fuentes et al. 2012; Kashif et al. 2012; Kreuze et al. 2013; Li et al. 2012; Loconsole et al. 2012a; Untiveros et al. 2010), negative- (Bi et al. 2012) and double- (Sela et al. 2012) stranded RNA genomes, and viroids (Li et al. 2012; Wu et al. 2012).

Whereas these studies indicate the broad applicability of sRSA among virus species, they are still relatively limited in the range of plant species to which sRSA has been applied (principally dicotyledonous plants) and a more extended analysis across different taxonomic groups of the plant kingdom would be beneficial.

General Lessons Regarding sRSA: Analysis of siRNA Sequence Datasets from Across the Plant Kingdom

To explore the efficiency of sRSA to identify viruses on a broad range of plant species we examined published plant small RNA sequence datasets available on the internet. The website of the project "comparative sequencing of plant small RNAs" at http://smallrna.udel.edu/ contains siRNA sequence data from plants throughout the plant kingdom (Table 1). We included into this analysis also three samples of

Table 1 Virus like sequences identified using siRNA datasets from 36 different species

	Source plant (common name)	Tissue	Contigs found (# and minimum e-value)[a]	Assembly[b]
Eudicots	Cucurbita maxima (pumpkin)	Leaves, phloem sap	–	–
	Populus trichocarpa (poplar)	Leaves	Caulimovirid (6; ≥9e-11)	54–144
	Gossypium arboretum (tree cotton)	Leaves, flowers, boll fiber	Totivirus (12, 0, 3; ≥5e-17)	51–204
	Citrus sinensis (sweet orange)	Leaves	Caulimovirid (11; ≥3e-07)	54–144
	Cairica papaya (papaya) [PRSV][c]	Leaves	PRSV	99.6 %
		Leaves	Partitivirus (8, 9, 0; ≥7e-19)	57–210
		Flowers	Totivirus (2, 0, 3; ≥2e-5)	54–99
	Silene latifolium (white campion)	Smutted flowers	Chrysovirus (0, 0, 6; ≥1e-10)[d]	67–198
			Caulimovirid (2, 4, 0; ≥4e-12)	57–102
	Physalis floridiana (Physalis)[e]	Leaves	TVCV-like (4; ≥0.012)	32–50
	Petunia hybrida (petunia)	Leaves	–	–
	Nicotiana tabaccum (tobacco)	Leaves	TVCV-like (23; ≥2e-10)	42–198
	N. benthamiana [SB-29][d,e]	Leaves	Torrado-like virus (5; ≥6e-5)	59–119
	Capsicum annum (sweet pepper)	Leaves, flowers, fruits	BPeV (29, 6, 9; ≥1e-35)	48–267
			TVCV-like (29, 11, 20; ≥5e-14)	42–240
	Solanum lycopersicum (tomato)	Leaves	TVCV-like (10; ≥4e-8)	45–111
	S. tuberosum (potato) [PVT][d,e]	Leaves	PVT	98.6 %
			TVCV-like (9; ≥4e-8)	42–105
	Lactuca sativa (lettuce)	Leaves, B. lactucea inoculated leaves, flowers	LBVaV RNA1-2 (56, 0, 3)	97.2 % (96.4–98)
			MiLV RNA1-4 (33, 0, 2)	93.3 % (85.6–99.6)
			BPYV (0, 0, 42)	96.9 % (96.1–97.8)
			TICV (0, 0, 14)	86.4 % (87.8–84.9)
			Badnavirus (1, 2, 17; ≥ 4e-13)	

(continued)

Table 1 (continued)

	Source plant (common name)	Tissue	Contigs found (# and minimum e-value)[a]	Assembly[b]
	Mimulus guttatus (Spotted monkey flower)	Leaves	–	–
	Persea americana (avocado)	Leaves	–	–
Magnolid	Zostera marina (sea grass)	Leaves	–	–
Monocots	Musa Acuminata (banana)	Leaves	Badnavirus (251; ≥8e-31)	42–207
	Oryza sativa (rice)	Leaves	RRSV RNA1-10	99.1 % (98.5–99.7)
	Triticum aestivum (wheat)	Leaves	–	–
	Hordeum vulgare (barley)	Leaves	–	–
	Panicum virgatum (switch grass)	Leaves	–	–
	Setaria italica (foxtail millet)	Leaves	–	–
	Zea mays (maiz)	Leaves	–	–
	Sorghum bicolor (Sorghum)	Leaves	–	–
	Miscanthus giganteus (Miscanthus)	Leaves	–	–
	Nuphar advena (water lilly)	Leaves, roots, flowers	CMV RNA1-3	99.5 %[f]
			Cytorhabdovirus (27,9,9; ≥9e-21)	42–228
			Caulimovirid (23,16,8; ≥6e-33)	42–315
Basal angiosperms	Amborella trichopoda (amborella)	Leaves	Caulimovirid (15; ≥2e-23)	51–228
	Picea abies (Norway spruce)	Leaves	–	–
Gymnosperms	Ginko biloba (maidenhair tree)	Leaves	–	–
	Cycas rumphii (Cycas)	Leaves	Caulimovirid (1; ≥4e-4)	69
	Marsilea quadrifolia (European water clover)	Leaves, roots, draught treated leaves	TuYV	99.6 %[g]

(continued)

Table 1 (continued)

	Source plant (common name)	Tissue	Contigs found (# and minimum e-value)[a]	Assembly[b]
Non-seed plants	Physcomitrella patens		Caulimovirid (4; ≥1e-3)	60–102
	Chara coralline (common stonewort)	Thallus	Caulimovirid (12; ≥4e-6)	39–120
Algae	Chlamydomonas reinhardtii (chlamydo monas)	Cells	–	–
	Volvox carteri (volvox)	Cells	–	–

[a]Number of contigs found, when more than one number is provided they corresponds to various tissues in the same order as named in the tissue column. Number of contigs or e-value not given when near complete genomes were obtained. –: no virus detected
[b]Size range of contigs obtained or % coverage of complete genome (in case of multiple genome components, range of coverage is given between *brackets*)
[c]Plant deliberately infected with virus between *square brackets*
[d]Chrysovirus contigs were only found in smutted leaves, while other viruses were found in all tissues
[e]Data from our own study
[f]Excluding 16–150 nt and 81–90 nt missing from 5′ and 3′ ends respectively
[g]Excluding 673 nt deletion and ~100 nt missing from 5′ end

our own: healthy *Physalis foridiana*, *Potato virus T* (PVT, *Flexiviridae*: *Tepovirus*) infected potato (*Solanum tuberosum*), and *Nicotiana benthamiana* infected with an uncharacterized virus SB-29. siRNA sequences assembled by Velvet for each of these datasets were submitted to BLASTX against the GenBank database, limiting the search to sequences of viral origin. For samples where known viruses were identified, full genomes of the most related sequence were downloaded and used for guide strand mediated assembly using MAQ (Fig. 2). The sequences and e-values of hits with significant similarity to viruses were extracted and are summarized in Table 1. Contig sequences are available from https://research.cip.cgiar.org/confluence/display/cpx/RsA+Plant+Kingdom. Analysis of this data provided a number of general lessons regarding BLASTX search results against the virus sequence database.

Reverse Transcribing Viruses

In most cases contigs with low level similarity (E values > 0.1) to animal retro- and pararetroviruses were found. These are unlikely to be viruses, but rather may represent unspecific hits, or sequences corresponding to degenerated or

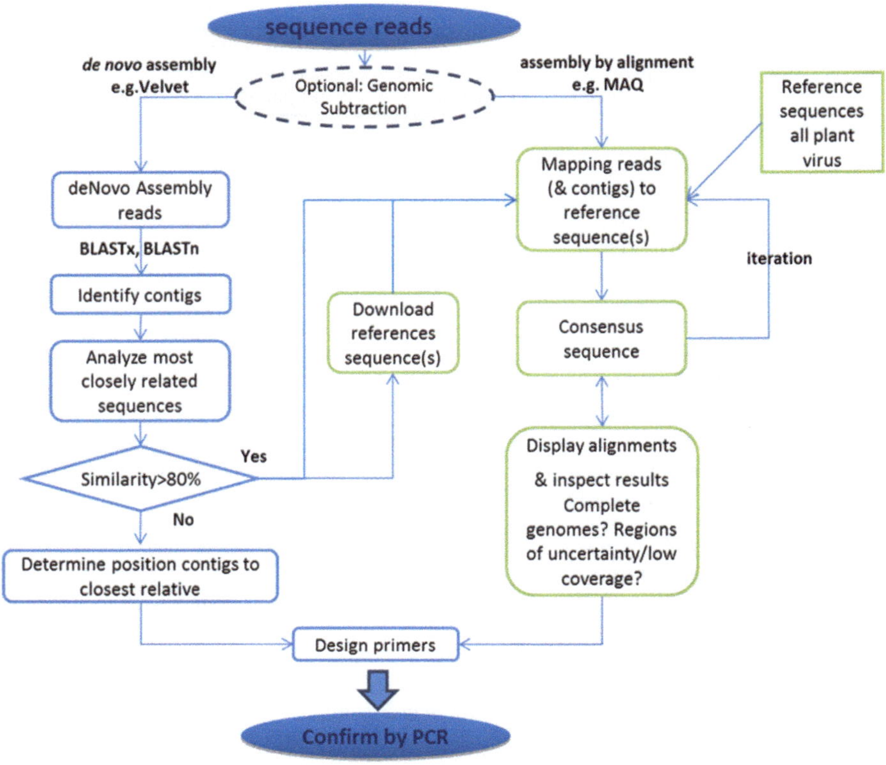

Fig. 2 General workflow for identifying viruses using sRSA performed in our lab

uncharacterized retro-elements in the plant genome. Furthermore, in many cases also contigs with significant similarity to viruses of the family *Caulimoviridae* were identified (Table 1). Although these may represent replicating viruses, they could also be the result of endogenous pararetrovirus like sequences or retrotransposons. Indeed, the contigs found in the moss *Physcomitrela patens* and tree *Populus trichocarpa* (Poplar) corresponded 100 % to genomic sequences when blasted against their respective genomes. Whereas in the case of *P. patens* the corresponding genomic fragment was annotated as an LTR retrotransposon (Genbank: GQ294565.1), the *P. trichocarpa* caulimovirus-like sequences corresponded to several predicted proteins, e.g. on scaffold_2615 (Genbank: NW_001490516.1), which contained two predicted ORFs corresponding to the coat protein and reverse transcriptase in the same gene order as found in viruses of the genera *Caulimovirus* and *Sobemovirus*. Numerous *Banana streak virus* (BSV, *Caulimoviridae*: *Badnavirus*) like sequences were identified in Banana (*Musa acuminata*, Table 1), a plant which is well known to contain endogenous BSV forms (Geering et al. 2005). In addition, contigs with high level of similarity to *Tobacco vein clearing virus* (TVCV; *Caulimoviridae*: *Cavemovirus*), were identified in all plants from the family *Solanaceae*, except petunia (*Petunia hybrida*).

Again, these are unlikely to represent replicating virus, but rather are genome integrated elements which have previously been shown to occur in potato (Hansen et al. 2005). This was confirmed by BLAST searches of the TVCV like contigs against the potato genome (www.potatogenome.net), which resulted in many hundreds of hits. Because TVCV like sequences have been reported also from other solanaceaous plants such as tomato (Staginnus et al. 2007) and various tobaccos (Lockhart et al. 2000; Gregor et al. 2004; Jakowitsch et al. 1999) these sequences may represent an ancient integration event maintained during the evolution within this plant family. Using primers designed to the contigs found in *P. floridana*, we were indeed able to amplify a 1,651 bp fragment encompassing partial ORF3 and ORF4 of a TVCV like sequence, which however contained numerous stop codons, suggesting it was indeed an inactive integrated sequence. On the other hand, no siRNA contigs with similarity to *Petunia vein clearing virus* (PVCV; *Caulimoviridae*: *Petuvirus*), or endogenous rice tungro bacilliform virus (ERTBV) were found in petunia or rice (*Oryza sativa*) respectively, although they have been shown to be integrated into genomes of these species (Kunii et al. 2004; Richert-Poggeler et al. 2003). This suggests that such sequences do not always lead to production of siRNAs, irrespective if they can be activated (PVCV) or not (ERTBV). Together these results show that pararetrovirus like sequences are common throughout the plant kingdom including algae, but that the identification of pararetrovirus like siRNA contigs does not unambiguously signify viral infection. Conversely, lack of pararetrovirus like contigs may not necessarily signify absence of integrated, or even activatable, pararetroviral sequences. Whereas the presence of stop codons in the translated reading frame of assembled contigs may suggest that inactive integrated viruses are present, it doesn't exclude extra-chromosomal viral replicons, and further analyses will be required to verify each particular case, particularly the first time a plant species is analyzed using this method. Availability of host genome sequences may help alleviate these problems through genomic subtraction (see section "Bioinformatics").

Known Viruses

Beyond contigs corresponding to reverse transcribing viruses, five new putative viruses as well as eight known viruses were identified among the 34 species downloaded (Table 1).

After identifying contigs with high similarity (>90 %) to known viruses in Papaya (*Papaya ring spot virus*, PRSV; *Potyviridae*: *Potyvirus*), Lettuce (*Mirafiori lettuce virus*, MiLV, *Ophioviridae*: *Ophiovirus*; *Lettuce big-vein associated virus*, LBVaV, *Varicosavirus*; *Beet pseudo yellows virus*, BPYV & *Lettuce infectious yellows virus*, LIYV, *Closteroviridae*: *Crinivirus*), Rice (*Rice ragged stunt virus*, RRSV; *Reoviridae*: *Oryzavirus*), Water lily (*Cucumber mosaic virus*, CMV, *Bromoviridae*: *Cucumovirus*), and European waterclover (*Turnip yellows virus*,

TuYV, *Luteoviridae*: *Polerovirus*), their complete sequences were downloaded and used for guide strand assisted assembly with MAQ and/or Novoalign. Nearly complete genome sequences could thus be assembled by combining *de novo* and guide strand mediated assembly (Table 1). It was notable that in most cases, assembly of 3′ and 5′ un-translated regions (UTRs) was poor and thus both UTRs of the CMV, MiLV and the 5′ UTR of TuYV were partly or entirely missing from the assemblies. Our experience is that viral UTRs are often, but not always, poor sources of siRNA as compared to coding regions and may reflect the evolution of these sequences to avoid degradation/targeting by RNA silencing. Also, in the multipartite viruses MiLV and RRSV, clear differences in coverage could be observed between different genome components, possibly reflecting their relative abundances in the plant. Whereas CMV, MiLV, LBVaV and RRSV were >98 % identical to published genomes, TuYV represented a new strain (Mq) of the virus as it had only ~90 % nt identity to the published sequence and also contained a 673 nt deletion eliminating the N-terminal 161 amino acids from the CP read-through cistron. The PRSV isolate showed the highest level of similarity (96 %) to isolates of Hawaii, where it was isolated (Emanuele De Paoli; personal communication). Comparison of the assemblies of the same virus from different tissues in the case of LBVaV, MiLV, CMV and TuYV, yielded practically (>99 %) identical consensus sequences where bases had been called, even though coverage varied significantly between tissues, e.g. the average number of unique siRNAs corresponding to TuYV were 18.9, 57.4 and 34 per 1,000 reads in leaf, root or draught stressed leaf samples, respectively.

New Viruses

Several new RNA viruses were also putatively predicted in some samples: a new totivirus from tree cotton, a new cytorhabdovirus from water lily, and three new viruses from white campion: another totivirus, a partitivirus and a chrysovirus. Chrysoviruses are typically fungal viruses and accordingly contigs with homology to chrysoviruses were only detected from smut infected flowers, and not healthy flowers or leaves, suggesting that the siRNAs originated from the smut pathogen (*Microbotryum violaceum*, Fuckel). Assuming this is the case, it implies that the method also works for fungi, as was previously predicted (Kreuze et al. 2009). The availability of data from different plant organs also enabled us to verify if the same viruses could be detected in different tissues. This was not always the case for the novel viruses when using BlastX searches of *de novo* assembled contigs (Table 1), even though MAQ was able to identify siRNAs homologous to viral contigs assembled from other tissues of the same plant, indicating virus specific siRNA were present in these tissues. This discrepancy is most likely due to that siRNAs were present at a too low frequency for efficient *de novo* assembly of contigs that were long enough to provide significant hits to the distantly related virus sequences available. Similar problems were indeed not experienced with the known viruses, which were readily detectable from any tissues with either BlastX or BlastN. The only exceptions were BPYV and LIYV, which could only be detected in the flower samples but they were the result of siRNA libraries taken from different plants (which were probably infected with additional viruses) than the leaf samples (Emanuele de Pauli, personal communication).

Although we are unable to confirm the results of this analysis because the data was generated by other scientist, and the original samples were not available to us, the analysis has been useful in highlighting both the scope of its applicability among plant species and tissue types, but also the challenges that are still faced, i.e., dealing with short contigs resulting from low coverage and how to distinguish integrated sequences from replicating viruses. The following section explores some of these limitations and the approaches that have been developed to improve them.

Factors Affecting the Efficacy of sRSA

Several factors can influence the efficiency of sRSA, and range from the sample preparation method to issues involving the bioinformatics analysis. Effects of sample preparation include method of RNA isolation, sample bulking and the siRNA cloning method. Although total RNA extraction kits can be used (Roy et al. 2013), most standard column based RNA extraction methods do not efficiently isolate small RNAs. Specialized kits for isolating miRNA can be used (Hwang et al. 2013; Sela et al. 2012), but alternatives such as Trizol (Kashif et al. 2012; Bi et al. 2012; Kreuze et al. 2009; Untiveros et al. 2010; De Souza et al. 2013; Fuentes et al. 2012) and CTAB (our lab, unpublished) protocols followed by gel purification of sRNAs have both proven to be adequate depending on the plant species and tissue type to be used. siRNA preparation steps prior to sequencing, which include sequential steps of adapter ligations and PCR enrichment can be fully outsourced to sequencing providers or performed in-house to save costs (e.g., Chen et al. 2012).

Effect of siRNA Cloning Method on Sequence Distribution and Assembly

Analysis of data from our first sRSA experiments (Kreuze et al. 2009) indicated that siRNA coverage of viral genomes was very unevenly distributed with short regions of extremely high coverage separated by regions with relatively low coverage (Fig. 3B). Linsen et al. (2009) showed that the method of siRNA library preparation could introduce a strong and consistent bias towards certain sequences depending on the method used and we suspected this was the case for our data as well. Uneven distribution of reads could potentially inhibit efficient *de novo* assembly of siRNAs.

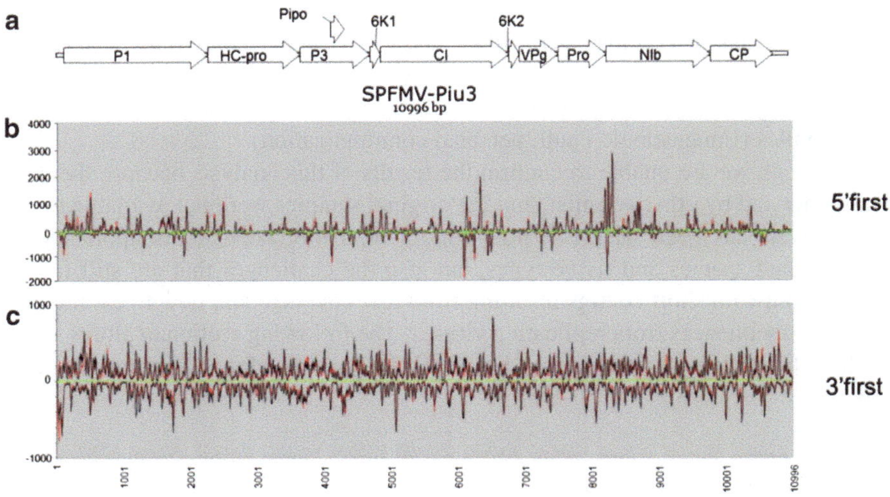

Fig. 3 Coverage of the sweet potato feathery mottle virus (SPFMV) genome by siRNAs following Illumina sequencing of the same samples using two different sample preparation techniques. (**a**) SPFMV genome structure. (**b**) Graph showing coverage for a library prepared by ligating the 5′ adaptor first. (**c**) Graph showing the coverage for a library prepared by ligating the 3′ adaptor first. The *lines* above and below the graph indicate the coverage of siRNAs corresponding to the positive or negative strand of the virus. The *different colors* of the *lines* indicate the different siRNA sizes: 21 nt *black*; 22 nt *red*; 23 nt *green*; 24 nt *yellow*

Therefore we decided to explore the effect of reversing the order of RNA adapter ligation in the sample preparation process of the exact same RNA extract from virus infected sweetpotato used previously (Kreuze et al. 2009). Whereas in the previous study the 5′ adapter was ligated first followed by the 3′ adapter, this time the 3′ adapter was ligated first followed by the 5′ adapter. Results showed that this simple change of procedure resulted in a much more homogenous distribution of siRNA reads over all five viral genomes analyzed and in all three samples (Fig. 3). Indeed Velvet was able to generate much larger viral contigs from each of the three samples and was able to assemble the whole genome of SPFMV without the need for manual re-assembly.

siRNA library preparation protocols have since been improved and currently recommended protocols are probably close to optimum. Nevertheless improvements may still be possible through adaptor and primer design, and should be kept in mind when choosing or developing in-house protocols.

Mixed Infections, Defective RNA/DNAs and Contamination

To make cost effective use of the current high throughput of NGS sequencers, samples are bulked and may be individually tagged by including a short DNA "barcode" or index during sample preparation. After a sequencing run each sample can then be disaggregated from the bulk based on their DNA index using computational algorithms. Care must be taken when preparing multiple samples as experience has shown the risk of contamination is significant and even minute amounts of cross sample contaminations can lead to assembly of significant portions of a virus genome, albeit mostly at low coverage. If siRNA libraries are prepared in-house it is recommended to always leave two empty lanes between each sample when running gel purifications. Still a recent study suggests 'virtual contamination' between indexed samples may occur at a rate of about 0.3 % during the computational processing of indexed libraries, and suggest double indexing as an efficient solution (Kircher et al. 2012).

Besides the issue of physical or virtual contaminations our experiences with sweetpotato field samples are that mixed infection with related viral strains are common. This is easily recognized by the assembly of contigs of the same region but differing in nucleotide sequence. Whereas this doesn't prevent virus identification, it makes assembly of reliable full genomes of each individual strain extremely difficult. Particularly when stretches of more than 20 nt with perfect identity exist between the two sequences, it may be difficult to exclude *in silico* recombination. Another confounding factor may be the presence of defective interfering (DI) DNA or RNA molecules, which usually replicate to much higher levels than the virus itself and produce large amounts of siRNA (Wu et al. 2010), which can confuse assembly software.

Bioinformatics

From the beginning, principally two different bioinformatics approaches have been followed to assemble viral genomes from sRNA sequences: *de novo* assembly of siRNAs (e.g. using Velvet), and alignment/mapping of sequences (e.g. with MAQ, BWA, NovoAlign). Often, the two are used in combination in an iterative process of de novo assembly and alignment to come to a final consensus (Fig. 2). Nevertheless, full or near complete virus genomes are often only achieved by assembly of siRNAs and or partial, *de novo* assembled, contigs onto a closely related reference genome. Vodovar et al. (2011), using *Flock house virus* as a model, developed a script, named Paparazzi, that reliably reconstitutes viral genomes through an iterative alignment/consensus call procedure using a related reference sequence as a scaffold. Still when nucleotide similarity drops below ~80 % efficiency and reliability of alignments of siRNAs are significantly reduced. Thus *de novo* assembly of siRNAs to achieve a complete genome for viruses where reference genomes with sufficient similarity are not available can be difficult.

One of the reasons limiting the ability of short sequence assemblers to generate full virus genomes de novo, may be interference in the assembly process by endogenous small RNA sequences, which are abundant and extremely polymorphic, often outnumbering viral siRNAs by orders of magnitude. Because mammalian antiviral siRNA are much less abundant than other eukaryotic organisms Isakov et al. (2011) used sRSA together with the concept of short RNA subtraction (or genomic subtraction) to address the problem of interference by host sRNAs. The method boosts pathogen-related siRNA signal through subtraction of host-derived sRNAs and led to successful identification of HIV in infected human cell cultures. Li et al. (2012) applied genomic subtraction to identify *Pepino mosaic virus* (*Potyviridae*: *Potyvirus*) and *Potato spindle tuber viroid* (*Pospiviroidae*: *Pospiviroid*) in tomato which resulted in a several-fold increase in the length of viral contigs and large reduction in the number of non-viral assembled contigs, as compared to assembly without genomic subtraction. Our own unpublished experiences with various viruses in potato also confirmed the increased assembly efficiency resulting from genomic subtraction, and it therefor seems recommendable to apply genomic subtraction whenever a genome sequence is available. However, care needs to be taken when perform genomic subtraction as some plant genome assemblies contain sequences of viral contaminations, besides endogenous integrated virus sequences.

One of the limitations for detecting new viruses is the need for sequences with a minimum level of homology in the database to achieve recognition. BlastX is efficient at identifying more distantly related sequences, but the success of identification of distantly related sequences will rapidly decrease with reduced contig size and for non-coding regions such as 5′ and 3′ UTRs, highlighting the benefit of assembling large contigs *de novo*. Still, considering the genetic diversity of known viruses, it is likely that there are still undiscovered viruses that exhibit no similarity to any of the currently known genera and families and these would escape detection by the current homology-dependent approaches. Genomic subtraction addresses this limitation to some extent, because any sequences remaining after subtraction of the host genome should automatically be considered as suspicious. One must however keep in mind that genomic subtraction should not be 100 % efficient, as the genomes of plant genotypes being analyzed may vary significantly from published genomes, particularly in heterozygous and polyploidy crops. For example, in potato the efficiency of genomic subtraction using the DM potato genome is only about 40–50 % when applied to other genotypes (nevertheless an improvement in *de novo* assembly of viral contigs is often observed). In addition, due to the high complexity of plant genome especially in centromere regions, currently no plant genomes have been sequenced and assembled to 100 % coverage. Wu et al. (2012) took another route to address the problem of homology requirement by developing a homology independent computational algorithm to identify replicating circular RNAs (referred to as progressive filtering of overlapping small RNAs: PFOR) and demonstrated its efficiency in identifying known viroids, but also a new viroid with no significant similarity to any previously known molecules. This approach seems promising, not only to identify viroids, but potentially also unknown circular DNA viruses. However, this approach requires the complete (100 %) coverage of the virus genome by siRNAs to ensure the identification of the circular structure of the genome and in our hands we were unable to recover the sequences of *Grapevine yellow speckle viroid 1* (*Pospiviroidae*: *Apscaviroid*) and *Hop stunt viroid*

(*Pospiviroidae*: *Hostuviroid*) from an infected grapevine, or *Sweet potato leaf curl virus* (*Geminiviridae*: *Begomovirus*) from infected sweetpotato plants using PFOR, whereas they were readily identified by assembly and BLAST as well as alignment with MAQ.

Concluding Remarks

Evidenced by the number of reports during the past several years sRSA has proven its power to identify and sequence diverse sets of viral and sub-viral agents including many new ones. The mining of published plant siRNA datasets presented in this chapter further expand on that, confirming its broad applicability but also revealing the remaining challenges. For sRSA to emerge from the research realm and become a mainstream generic viral detection method however important improvements are still required, particularly on the side of the sequence analysis. Whereas different bioinformatics solutions have been presented to address specific situations, they don't yet present a unified solution addressing each of the different issues. Development of automated and universally applicable bioinformatics pipelines for analysis of sRNA data, specifically for virus identification, should be a priority for future development and are required before the method can become part of regular viral diagnostics.

Acknowledgments I acknowledge the contributions from the various present and previous members of my lab and colleagues at CIP, and Emanuele De Paoli and Manfred Ruddat for sharing information about the samples from their small RNA libraries. Sincere thanks also to Zhangjun Fei for critical reading and useful comments of this chapter. The work on sRSA in our lab has been funded by grants from the Howard Buffet Foundation (subgrant 20127-C), Bill and Melinda Gates Foundation through the SASHA project (Grant OPP53344), the National Science Foundation through its BREAD program (Award No. 1110080), ICGEB-TWAS-UNESCO/IBSP Joint Programme on Capacity Building in Basic Molecular Biology (Contract No.: CRP/101005) and EU FP7 through the QBOL project (Grant No. 226482).

References

Adams IP, Glover RH, Monger WA, Mumford R, Jackeviciene E, Navalinskiene M, Samuitiene M, Boonham N (2009) Next-generation sequencing and metagenomic analysis: a universal diagnostic tool in plant virology. Mol Plant Pathol 10(4):537–545. doi:10.1111/j.1364-3703.2009.00545.x

Adams IP, Miano DW, Kinyua ZM, Wangai A, Kimani E, Phiri N, Reeder R, Harju V, Glover R, Hany U, Souza-Richards R, Nath PD, Nixon T, Fox A, Barnes A, Smith J, Skelton A, Thwaites R, Mumford R, Boonham N (2013) Use of next-generation sequencing for the identification and characterization of maize chlorotic mottle virus and sugarcane mosaic virus causing maize lethal necrosis in Kenya. Plant Pathol 62(4):741–749. doi:10.1111/j.1365-3059.2012.02690.x

Al Rwahnih M, Daubert S, Golino D, Rowhani A (2009) Deep sequencing analysis of RNAs from a grapevine showing Syrah decline symptoms reveals a multiple virus infection that includes a novel virus. Virology 387(2):395–401. doi:10.1016/j.virol.2009.02.028

Al Rwahnih M, Daubert S, Urbez-Torres JR, Cordero F, Rowhani A (2011) Deep sequencing evidence from single grapevine plants reveals a virome dominated by mycoviruses. Arch Virol 156(3):397–403. doi:10.1007/s00705-010-0869-8

Bi Y, Tugume AK, Valkonen JP (2012) Small-RNA deep sequencing reveals Arctium tomentosum as a natural host of Alstroemeria virus X and a new putative Emaravirus. PLoS One 7(8):e42758. doi:10.1371/journal.pone.0042758

Candresse T, Marais A, Faure C, Gentit P (2013) Association of little cherry virus 1 (LChV1) with the Shirofugen stunt disease and characterization of the genome of a divergent LChV1 isolate. Phytopathology 103(3):293–298. doi:10.1094/phyto-10-12-0275-r

Chen Y-R, Zheng Y, Liu B, Zhong S, Giovannoni J, Fei Z (2012) A cost-effective method for Illumina small RNA-Seq library preparation using T4 RNA ligase 1 adenylated adapters. Plant Methods 8:4

Coetzee B, Freeborough MJ, Maree HJ, Celton JM, Rees DJ, Burger JT (2010) Deep sequencing analysis of viruses infecting grapevines: virome of a vineyard. Virology 400(2):157–163. doi:10.1016/j.virol.2010.01.023

Cuellar WJ, Cruzado RK, Fuentes S, Untiveros M, Soto M, Kreuze JF (2011) Sequence characterization of a Peruvian isolate of sweet potato chlorotic stunt virus: further variability and a model for p22 acquisition. Virus Res 157(1):111–115. doi:10.1016/j.virusres.2011.01.010

De Souza J, Fuentes S, Savenkov S, Cuellar W, Kreuze J (2013) The complete nucleotide sequence of sweet potato C6 virus: a carlavirus lacking a cysteine-rich protein. Arch Virol 158 (6):1393–1396. doi:10.1007/s00705-013-1614-x

Fuentes S, Heider B, Tasso RC, Romero E, Zum Felde T, Kreuze JF (2012) Complete genome sequence of a potyvirus infecting yam beans (*Pachyrhizus* spp.) in Peru. Arch Virol 157 (4):773–776. doi:10.1007/s00705-011-1214-6

Geering AD, Olszewski NE, Harper G, Lockhart BE, Hull R, Thomas JE (2005) Banana contains a diverse array of endogenous badnaviruses. J Gen Virol 86(Pt 2):511–520. doi:10.1099/vir.0.80261-0

Gregor W, Mette MF, Staginnus C, Matzke MA, Matzke AJM (2004) A distinct endogenous pararetrovirus family in Nicotiana tomentosiformis, a diploid progenitor of polyploid tobacco. Plant Physiol 134(3):1191–1199. doi:10.1104/pp. 103.031112

Haasnoot J, Westerhout EM, Berkhout B (2007) RNA interference against viruses: strike and counterstrike. Nat Biotechnol 12:1435–1443

Hagen C, Frizzi A, Gabriels S, Huang M, Salati R, Gabor B, Huang S (2012) Accurate and sensitive diagnosis of geminiviruses through enrichment, high-throughput sequencing and automated sequence identification. Arch Virol 157(5):907–915. doi:10.1007/s00705-012-1253-7

Hansen CN, Harper G, Heslop-Harrison JS (2005) Characterisation of pararetrovirus-like sequences in the genome of potato (Solanum tuberosum). Cytogenet Genome Res 110(1–4):559–565. doi:10.1159/000084989

Huang Y, Mi ZQ, Zhuang L, Ma MJ, An XP, Liu W, Cao WC, Tong YG (2013) Presence of entomobirnaviruses in Chinese mosquitoes in the absence of Dengue virus co-infection. J Gen Virol 94:663–667. doi:10.1099/vir.0.048231-0

Hwang YT, Kalischuk M, Fusaro AF, Waterhouse PM, Kawchuk L (2013) Small RNA sequencing of potato leafroll virus-infected plants reveals an additional subgenomic RNA encoding a sequence-specific RNA-binding protein. Virology 438(2):61–69. doi:10.1016/j.virol.2012.12.012

Isakov O, Modai S, Shomron N (2011) Pathogen detection using short-RNA deep sequencing subtraction and assembly. Bioinformatics 27(15):2027–2030. doi:10.1093/bioinformatics/btr349

Jakowitsch J, Mette MF, van der Winden J, Matzke MA, Matzke AJM (1999) Integrated pararetroviral sequences define a unique class of dispersed repetitive DNA in plants. Proc Natl Acad Sci U S A 96(23):13241–13246. doi:10.1073/pnas.96.23.13241

Kashif M, Pietilä S, Artola K, Jones RAC, Tugume AK, Mäkinen V, Valkonen JPT (2012) Detection of viruses in sweetpotato from Honduras and Guatemala augmented by deep-sequencing of small-RNAs. Plant Dis 96(10):1430–1437. doi:10.1094/pdis-03-12-0268-re

Kircher M, Sawyer S, Meyer M (2012) Double indexing overcomes inaccuracies in multiplex sequencing on the Illumina platform. Nucleic Acids Res 40(1):e3. doi:10.1093/nar/gkr771

Kreuze JF, Pérez A, Untiveros M, Quispe D, Fuentes S, Barker I, Simon R (2009) Complete viral genome sequence and discovery of novel viruses by deep sequencing of small RNAs: a generic method for diagnosis, discovery and sequencing of viruses. Virology 388(1):1–7. doi:10.1016/j.virol.2009.03.024

Kreuze J, Koenig R, De Souza J, Vetten HJ, Muller G, Flores B, Ziebell H, Cuellar W (2013) The complete genome sequences of a Peruvian and a Colombian isolate of Andean potato latent virus and partial sequences of further isolates suggest the existence of two distinct potato-infecting tymovirus species. Virus Res 173(2):431–435

Kunii M, Kanda M, Nagano H, Uyeda I, Kishima Y, Sano Y (2004) Reconstruction of putative DNA virus from endogenous rice tungro bacilliform virus-like sequences in the rice genome: implications for integration and evolution. BMC Genomics 5(10):80. doi:10.1186/1471-2164-5-80

Li R, Gao S, Hernandez AG, Wechter WP, Fei Z, Ling KS (2012) Deep sequencing of small RNAs in tomato for virus and viroid identification and strain differentiation. PLoS One 7(5):e37127. doi:10.1371/journal.pone.0037127

Linsen SE, de Wit E, Janssens G, Heater S, Chapman L, Parkin RK, Fritz B, Wyman SK, de Bruijn E, Voest EE, Kuersten S, Tewari M, Cuppen E (2009) Limitations and possibilities of small RNA digital gene expression profiling. Nat Methods 6(7):474–476. doi:10.1038/nmeth0709-474

Lockhart BE, Menke J, Dahal G, Olszewski NE (2000) Characterization and genomic analysis of tobacco vein clearing virus, a plant pararetrovirus that is transmitted vertically and related to sequences integrated in the host genome. J Gen Virol 81:1579–1585

Loconsole G, Onelge N, Potere O, Giampetruzzi A, Bozan O, Satar S, De Stradis A, Savino V, Yokomi RK, Saponari M (2012a) Identification and characterization of citrus yellow vein clearing virus, a putative new member of the genus Mandarivirus. Phytopathology 102 (12):1168–1175. doi:10.1094/PHYTO-06-12-0140-R

Loconsole G, Saldarelli P, Doddapaneni H, Savino V, Martelli GP, Saponari M (2012b) Identification of a single-stranded DNA virus associated with citrus chlorotic dwarf disease, a new member in the family Geminiviridae. Virology 432(1):162–172. doi:10.1016/j.virol.2012.06.005

Ma M, Huang Y, Gong Z, Zhuang L, Li C, Yang H, Tong Y, Liu W, Cao W (2011) Discovery of DNA viruses in wild-caught mosquitoes using small RNA high throughput sequencing. PLoS One 6(9):e24758. doi:10.1371/journal.pone.0024758.t001

Mlotshwa S, Pruss GJ, Vance V (2008) Small RNAs in viral infection and host defense. Trends Plant Sci 13(7):375–382

Ng TFF, Duffy S, Polston JE, Bixby E, Vallad GE, Breitbart M (2011) Exploring the diversity of plant DNA viruses and their satellites using vector-enabled metagenomics on whiteflies. PLoS One 6(4):e19050. doi:10.1371/journal.pone.0019050

Richert-Poggeler KR, Noreen F, Schwarzacher T, Harper G, Hohn T (2003) Induction of infectious petunia vein clearing (pararetro) virus from endogenous provirus in petunia. EMBO J 22 (18):4836–4845. doi:10.1093/emboj/cdg443

Roossinck MJ, Saha P, Wiley GB, Quan J, White JD, Lai H, Chavarría F, Shen G, Roe BA (2010) Ecogenomics: using massively parallel pyrosequencing to understand virus ecology. Mol Ecol 19(1):81–88

Roy A, Choudhary N, Guillermo LM, Shao J, Govindarajulu A, Achor D, Wei G, Picton DD, Levy L, Nakhla MK, Hartung JS, Brlansky RH (2013) A novel virus of the genus Cilevirus causing symptoms similar to *Citrus leprosis*. Phytopathology 103(5):488–500. doi:10.1094/phyto-07-12-0177-r

Sela N, Luria N, Dombrovsky A (2012) Genome assembly of bell pepper endornavirus from small RNA. J Virol 86(14):7721–7721. doi:10.1128/jvi.00983-12

Staginnus C, Gregor W, Mette MF, Teo CH, Borroto-Fernandez EG, Machado ML, Matzke M, Schwarzacher T (2007) Endogenous pararetroviral sequences in tomato (Solanum lycopersicum) and related species. BMC Plant Biol 7:24. doi:10.1186/1471-2229-7-24

Untiveros M, Quispe D, Kreuze J (2010) Analysis of complete genomic sequences of isolates of the *Sweet potato feathery mottle virus* strains C and EA: molecular evidence for two distinct potyvirus species and two P1 protein domains. Arch Virol 155(12):2059–2063. doi:10.1007/s00705-010-0805-y

Vodovar N, Goic B, Blanc H, Saleh MC (2011) In silico reconstruction of viral genomes from small RNAs improves virus-derived small interfering RNA profiling. J Virol 85 (21):11016–11021. doi:10.1128/JVI.05647-11

Wu Q, Luo Y, Lu R, Lau N, Lai EC, Li W-X, Ding S-W (2010) Virus discovery by deep sequencing and assembly of virus-derived small silencing RNAs. Proc Natl Acad Sci U S A 107(4):1606–1611. doi:10.1073/pnas.0911353107

Wu Q, Wang Y, Cao M, Pantaleo V, Burgyan J, Li W-X, Ding S-W (2012) Homology-independent discovery of replicating pathogenic circular RNAs by deep sequencing and a new computational algorithm. Proc Natl Acad Sci U S A 109(10):3938–3943. doi:10.1073/pnas.1117815109

Wylie SJ, Jones MGK (2011) The complete genome sequence of a Passion fruit woodiness virus isolate from Australia determined using deep sequencing, and its relationship to other potyviruses. Arch Virol 156(3):479–482. doi:10.1007/s00705-010-0845-3

Wylie SJ, Luo H, Li H, Jones MGK (2012a) Multiple polyadenylated RNA viruses detected in pooled cultivated and wild plant samples. Arch Virol 157(2):271–284. doi:10.1007/s00705-011-1166-x

Wylie SJ, Tan AJY, Li H, Dixon KW, Jones MGK (2012b) Caladenia virus A, an unusual new member of the family Potyviridae from terrestrial orchids in Western Australia. Arch Virol 157 (12):2447–2452. doi:10.1007/s00705-012-1452-2

Wylie SJ, Li H, Dixon KW, Richards H, Jones MGK (2013) Exotic and indigenous viruses infect wild populations and captive collections of temperate terrestrial orchids (Diuris species) in Australia. Virus Res 171(1):22–32. doi:10.1016/j.virusres.2012.10.003

Zhang Y, Singh K, Kaur R, Qiu W (2011) Association of a novel DNA virus with the grapevine vein-clearing and vine decline syndrome. Phytopathology 101(9):1081–1090. doi:10.1094/PHYTO-02-11-0034

Use of Airborne Inoculum Detection for Disease Management Decisions

Walter F. Mahaffee

Abstract Knowledge of inoculum presence has been used for decades to help guide disease management decisions. However, its implementation on a broad scale has been limited due to the capital costs and technical skill required to effectively monitor pathogen presence across large areas. Recent advances in nucleic acid detection technologies are showing promise in enabling field level implementation of inoculum detection and quantification to aid in disease management decisions. Over the past 7 years we have investigated the use various molecular approaches to monitor the presence of *Erysiphe necator* in commercial vineyards and time fungicide applications based on its detection. We have found loop mediated isothermal amplification (LAMP) to be a robust method for the detection of *E. necator* DNA that may be suitable for practitioner implementation.

Keywords LAMP • qPCR • Inoculum detection • Airbiotia • Spore dispersion • IPM

Introduction

Monitoring for inoculum presence to aid in disease management decisions is not a new concept and has been used for decades in numerous pathosystems (Campbell and Madden 1990; Dhingra and Sinclair 1995). The majority of these approaches are related to determining the amount of initial inoculum for soil borne diseases; however, there have been several successes in monitoring airborne inoculum to aid in disease management systems. In the 1960s, Berger (1969) developed a disease

W.F. Mahaffee (✉)
United States Department of Agriculture – Agricultural Research Service, Horticultural Crops Research Unit, Corvallis, OR 97330, USA

Department of Botany and Plant Pathology, Oregon State University, Corvallis, OR 97331, USA
e-mail: mahaffew@science.oregonstate.edu

M.L. Gullino and P.J.M. Bonants (eds.), *Detection and Diagnostics of Plant Pathogens*, Plant Pathology in the 21st Century 5, DOI 10.1007/978-94-017-9020-8_3,

forecasting model for celery early blight that was based on the number of *Cerospora apii* spores trapped and consecutive hours above 15 °C. Harrison et al. (1965) used presence of airborne inoculum to initiate fungicide application to manage potato early blight caused by *Alternaria solani*. Although neither of these approaches is still used, they were successful in reducing fungicide usage without compromising disease control. These studies demonstrate that knowledge of inoculum availability can aid management of airborne plant pathogens. Currently, the hop downy mildew disease forecaster developed by Royle (Royle 1973; Kremheller and Diercks 1983) is used in the Hallertau region of Germany to time fungicide applications. This system relies on a combination of weather based disease forecasting and the visual identification and quantification of *Psuedoperonospora humuli* sporangia to guide timing of fungicide applications; thus, demonstrating that monitoring airborne inoculum can be commercially implemented.

While inoculum monitoring can be useful for aiding in disease management decision, it has always been difficult to implement on a broad scale due to the difficulty in and cost of sample collection and visual identification of infective propagules (McCartney et al. 1997; West et al. 2008). To reduce the time required for assessing samples and increase confidence in inoculum identification various immunological (Kennedy et al. 2000; De Boer and López 2012) and nucleic acid based technologies have been developed that are suitable for detecting and quantifying airborne inoculum (Calderon et al. 2002; Falacy et al. 2007; Gent et al. 2009; Carisse et al. 2012).

Implementation of most nucleic acid detection technologies on a board scale in agriculture is still limited due to the capital cost and technical expertise required to insure accuracy of the results. Recent advancements in nucleic acid amplification using non-PCR approaches have resulted in techniques that show promise in enabling field practitioners to perform the assay and use knowledge of inoculum presence and/or concentration to inform management decisions (e.g. initiate and time fungicide applications).

Epidemiological Concepts for Monitoring Airborne Inoculum

Most disease management strategies of airborne plant pathogens in practice are based on the assumption that inoculum is always present (Hardwick 2006) and often fail to predict the severity of the epidemic because they do not account for the quantity of initial inoculum present at a location (West et al. 2008). There are numerous reasons for differences in initial inoculum including differences in microclimate or management practices that impact inoculum survival or the amount formed the previous season. For example, with grape powdery mildew cleistothecia are considered the predominant overwintering structure and are formed in the late summer to early fall (Pearson and Gadoury 1987; Rossi et al. 2010). The amount formed is considered to be a function of the disease severity as influenced by canopy density and microclimate, which result in the aggregation of overwintering inoculum (Mahaffee, unpublished). Similar aggregation of bud perennation by *E. necator* has been observed in grape (Cortesi et al. 2005).

Another concern associated with predicting epidemic development is that many economically important pathosystems involves either an exotic (not native to the growing region) pathogen or host and many times both. Since the pathogen and/or host did not evolve in the environment where they occur, there are likely biological limitations that impact pathogen survival or an asynchrony in the host and pathogen life cycles that impacts the availability of inoculum during periods of host susceptibility. This asynchrony can affect the reliability of disease forecasters because many are only initiated once susceptible tissue is present (e.g. planting for annuals or bud burst for perennials) (Gubler et al. 1999). In fact, inoculum avoidance is a common practice in some agricultural systems (Jacobsen 2007) and could be naturally achieved in some regions if the asynchrony between the host and pathogen is large enough.

Two classic examples of the phenomenon of limited pathogen overwintering is the annual tobacco blue mold and cereal rust epidemics in the United States (Nagarajan and Singh 1990; Aylor 2003). *Peronospora tabacina* (tobacco blue mold) is not known to overwinter above the 30th parallel in the western hemisphere. Each year the pathogen is thought to be transported in storm systems or other wind currents moving to the northern latitudes of North America from overwintering sites in the commercial fields in the Caribbean or wild *Nicotiana* spp. in the Southwest. A decision support system that predicts the potential of infection based on the weather forecast, potential of transport, and estimated source strength was developed to monitor and predict inoculum availability on a regional basis. The source strength is estimated from disease reports and transport forecasting that is augmented with disease monitoring at sentinel lots and grower fields. This information is then distributed to growers as alerts. Similar systems have been established for cucumber downy mildew (Ojiambo and Holmes 2011) and soybean rust (Hershman et al. 2011). These systems have been very effective in reducing fungicide usage and improving disease management across a production region but they require extensive resources and coordination that are not realistic for many cropping systems. They also fail to consider differences in microclimate or topography that will impact inoculum deposition and infection.

Assay Traits Required for Grower Performed Inoculum Detection

In order for any of these technologies to be suitable for practical use in monitoring of airborne plant pathogens, they must be exceptionally sensitive (e.g. less than 10 propagative units) and able to detect the target in an overwhelming background of non-target DNA (Alvarez et al. 1995; Fröhlich-Nowoisky et al. 2009) using crude DNA extraction techniques. Sensitivity can be especially problematic under these conditions since it is usually achieved by either reducing the liquid volume in which the DNA is extracted or by transferring large volumes of the DNA extraction into an assay master mix. Either approach results in a large amount of inhibitors or background DNA being transferred into the reaction. Thus, the master mix must be very robust and resistant to inhibitors and the polymerase and primers need to be at concentrations that will allow for rapid location of the target sequence in order for the reaction to proceed at a suitable speed. The assay must also be robust enough to handle a constantly changing complex of inhibitors that results from the highly variable particle composition in the air throughout a growing season (Tong and Lighthart 2000; Jaenicke 2005; Lee et al. 2006; Khattab and Levetin 2008).

Sample collection and processing must be relatively quick and require few operations in order to be consistently used by practitioners. Practitioners will have little patience or time to implement complex procedures. There is also an increased potential for a false result due to operator error or contamination with each step. For instance; LAMP reactions should never be opened post amplification because the stability and quantity of the amplified product results in a high potential for contamination of the work surface or equipment. Lastly, the results must be easy to assess either visually or automatically.

Methods for Monitoring Airborne Pathogen Inoculum

Perhaps the most common method for monitoring the presence of airborne pathogen inoculum is scouting for disease symptoms in a management unit. Some models use measurement of disease in the previous year to help predict disease development in the current growing season (Kast 2000), while others use measurements of early season disease incidence to initiate or adjust fungicide application intervals (Jacobi et al. 1995). The latter is very labor intensive since it is very difficult to accurately assess low levels of disease in large fields (Madden and Hughes 1999; Gent et al. 2007), particularly when disease is often aggregated in the early stages of epidemics (Cortesi et al. 2004; Turechek and Mahaffee 2004; Mundt 2009). There is also the issue of under estimating the amount of disease due to latency or lack of visual symptoms (chapter 3 of Madden et al. 2007). Furthermore, assessment of disease symptoms only estimates how much inoculum was present sometime prior (variable based on environment and latency of the pathogen) to the date of assessment. Thus, there is likely much more infected plant tissue in the field than is visible at the time of assessment (chapter 10 of Madden et al. 2007). Depending on the previous and current weather conditions, the amount of inoculum being generated, critical periods of host susceptibility and the frequency of scouting, this approach could lead to control failures (Magarey et al. 2002). These issues are the impetus for the research in other approaches to assess inoculum availability (e.g. detecting the presence of pathogen inoculum in the air).

The practical assessment of air borne inoculum presence requires a means of collecting airborne propagules that is both easily processed and inexpensive. There are two main approaches to sampling airborne inoculum: passive sampling and volumetric sampling.

Passive sampling relies on either gravitational forces to cause settling of airborne propagules to horizontal surfaces (e.g. coated glass slides, agar plates) in the area of interests or inertia to imping particles onto a vertical surface. Although quite cheap and easy to implement, the highly variable sample volume associated with passive sampling strategies (Gregory 1973a) limits their utility in monitoring for pathogen presence or quantity in most pathosystems. This approach also tends to utilize a large sampling surface which can be advantageous for visual detection but poses problems for other detection methods. The larger the surface area typical of this method requires large volume extractions that would likely increase extraction cost and time and reduce sensitivity due to dilution. This approach also requires monitoring of wind speed and direction to determine the air volume sampled.

Volumetric samplers utilize three main approaches (inertia, filtration, or cyclonic/centrifugal separation) to collect propagules by moving either known volumes of air over the sampling surface or by moving the sampling surface at a known rate through the air to cause impaction of airborne propagules onto or in sampling matrices; thus, achieving a standard air sample volume (Gregory 1973b). Electrostatic charge (Schneider et al. 2009) has also been used to collect airborne propagules onto a sampling matrix. Volumetric sampling has the advantage of accurate quantification of sample volume that is less impacted by ambient wind speed and direction (West et al. 2009). However, particle rebound from the sampling surface does occur, especially once the sampling surface becomes loaded with trapped particles. Some suction based impaction traps (e.g. Hirst, Burkhardt) have the added advantage of a moving sampling surface that allow for the relative time of collection to be determined in addition to reducing particle loss due overloading of sample surface. However, the solid surfaces employed tend to be large and continuous which may make processing difficult and reduce sensitivity (Calderon et al. 2002). They also tend to have lower sample volumes (10–20 L/min) that limit their utility in detection of the low concentrations of air-borne spores that are common early in disease epidemics (Aylor 1989a). Many are also cost prohibitive for practitioner level implementation. Traps based on a liquid sample matrix are advantageous for sample processing but have the disadvantage that propagule integrity is often compromised. Some spores will lyse while others may grow, which could impact detection and quantification. Traps based on movement of the sampling surface through the air are very efficient and inexpensive for collecting particles and are suitable for numerous identification methods. However, they tend to have a small sample surface that can easily become overloaded and reduce collection efficacy due to particle rebound.

Trapping systems based of *filtration* of air drawn through a porous or liquid matrix are advantageous since the sample is fairly easy to process and the traps can be inexpensive. However, porous filters have a tendency to rapidly clog when the air stream has a high particle load (e.g. agricultural environments) and large air volumes are sampled (Gregory 1973a). A liquid matrix can also loose samples to particles being released when bubbles burst at the liquid surface, releasing particles back into the air (Gregory 1973a). Care must also be taken with liquid sample matrices to reduce spore lysis or to preserve nucleic acids released when spore lysis occurs. Spore germination and growth might also be an issue if it causes and increase in the number of nuclei and enumeration is a goal.

Some impaction traps work by generating an air vortex that causes particles to be deposited onto the walls, bottom of a sampling vessel (Lacey and West 2006) or into liquid that runs down the sampler walls due to the centrifugal forces created by the vortex (e.g. BioGaurdian, Innovatek, Inc). These samplers differ greatly in their sample volumes ranging from 1 to 1,000 L/min but generally concentrate the spores into a small volume that can easily be processed for numerous detection methods. Their main disadvantage is the expense and that the required energy can be difficult to provide in agricultural settings.

Electrostatic samplers (Schneider et al. 2009) are a relatively new sampling method that could be very useful once perfected. The samples are relatively easy to process and can achieve very high sample volumes. The disadvantages can be the generation of ozone or electrical arcing that impacts propagule integrity and their robustness under harsh agricultural conditions. They also tend to require more energy than other traps but not as much as most cyclonic traps.

The land area represented by any of these trapping methods will be extremely dynamic due to the dependency on source strength, air turbulence and velocity, environmental conditions, local topology, host and pathogen traits, and trapping method (Gregory 1973a). The interaction among these factors results in a constantly changing airflow, thus the land area represented by a spore trap is in constant flux. Our experience indicates that placement of a trap in the turbulent mixing layer just above a grape canopy is likely the best location for monitoring the asexual phase of *E. necator* since traps placed in this region more consistently collected spores than traps placed on the ground or at the leeward edge (based on the predominate airflow direction) of a vineyard. Recently Van der Heyden et al. (2013) systematically examined trap placement to monitor strawberry powdery mildew and found significant heterogeneity at a spatial scale of 100 m^2 but that only one spore trap was needed to monitor a field (1,400 m^2) when action thresholds were used to inform management decisions. The assertion that the land mass represented by a trap is limited is also supported by epidemiological theory and data. Most dispersion kernels have an exponential decline in inoculum concentration (Aylor 1989b; Sackett and Mundt 2005; Schmale et al. 2005) which indicates that there is rapid and extreme dilution of propagules from a source. In a recent experiment examining the dispersion of microspheres (surrogate spores) in a commercial vineyard, we observed a ninefold decrease in particle concentration in 180 m (Fig. 1). These results and observations indicate that individual spore traps are not likely to represent large areas.

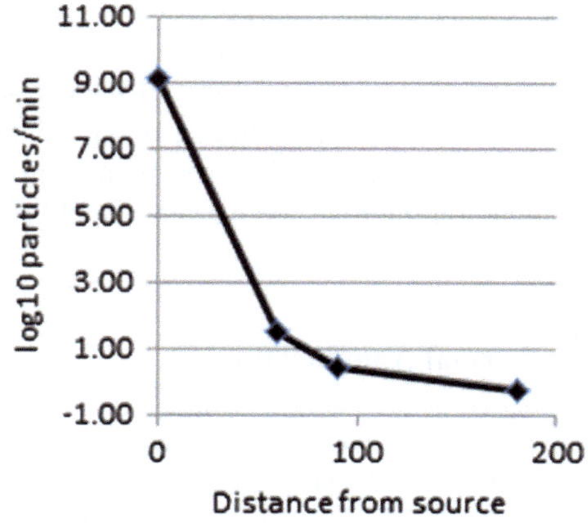

Fig. 1 Concentration of 10–40 μm microspheres in the air with distance from the source (Mahaffee et al. unpublished)

Methods for Practitioner Pathogen Identifications

Methods intended for use by practitioners must have several traits that ensure their use and effectiveness in achieving practitioner objectives. Assays must be simple to perform and yield results quickly and unambiguously. These goals are not easily achieved and often result in compromises in sensitivity or specificity (see De Boer and Lopez 2012 for a more detailed review).

Immunological Testing

The most common practitioner preformed detection assays are lateral flow immunoassays. They are cheap and rapid (<10 min) but tend to be insensitive (e.g. high false positive rate) (Posthuma-Trumple et al. 2009). Some recent work has paired nucleic acid amplification with lateral flow immune detection (Tomlinson et al. 2010). While these approaches improve the sensitivity of the later flow assays, they also increase the cost and complexity and in the case of some isothermal techniques increase to possibility of contaminated reactions.

Nucleic Acid Testing

Monitoring of airbiota using a PCR approach has been extensively investigated (West et al. 2008) and shows great promise in helping to understand pathogen biology and disease epidemiology. However, PCR still requires expensive equipment and skilled labor that limits its utility for use by most practitioners. Despite numerous attempts to develop field accessible PCR devices (Kuske et al. 1998; Schaad et al. 2003; Tomlinson et al. 2007), this approach still remains largely the purview of a lab. Transportation of samples from collection points to processing labs often results in significant time delays that reduce the utility of data for real-time management decisions. However, these techniques can give quantitative estimates of spore concentration (McDevitt et al. 2004; Carisse et al. 2009; Rogers et al. 2009) that can be used to determine when to extend or reduce fungicide application intervals (Mahaffee, unpublished).

Isothermal Amplification

In the past 15 years, several new isothermal nucleic amplification technologies have been developed that could be adapted for use by practitioners (Gill and Ghaemi 2008; Niemz et al. 2011; Craw and Balachandran 2012). These methods can be divided into five main categories based on method of amplification, replication, and amplification product.

RNA Transcription. These methods [e.g. transcription mediated amplification (TMA), nucleic acid sequence based amplification (NASBA), signal-mediated amplification of RNA technology (SMART)] utilize DNA and RNA polymerases to amplify target sequences. Since most inoculum monitoring methods will have extended periods where the propagules remain entrained on an inert matrix, RNA template based methods are likely to be unsuitable for monitoring inoculum availability due to the greater degradation potential of most RNA templates.

DNA Replication with Enzymatic Strand Duplex Denaturation. This group employs helicases or recombinases to enzymatically separate DNA strands and single strand binding proteins to stabilize single strands of DNA that are then amplified by DNA polymerase. Helicase dependent amplification (HDA) and recombinase polymerase amplification (RPA) are two common examples of this approach. These approaches have been shown to be effective across a wide range of sample types but are currently often insensitive (>100 copies of target) and require extended amplification times or multistep processing to reduce amplification time. There are numerous advancements in the medical community that could help make this technology appealing for use in monitoring airborne pathogens.

Nuclease Assisted Amplification. This method utilizes an initial heat denaturation followed by DNA polymerase amplification and nuclease activity. There are several different strategies that employ various combinations of primers and nuclease activity. Strand displacement (SDA) uses primers with sites for double-strand endonucleases, while nicking enzyme amplification reaction (NEAR), nicking enzyme-mediated amplification (NEMA), beacon-assisted detection amplification, and exponential amplification reaction (EXPAR) utilize single-strand endonucleases to nick DNA. Isothermal and Chimeric primer-initiated amplification of nucleic acids (ICAN) uses DNA-RNA-DNA chimer primers and RNAse activity to nick single strands and allow DNA polymerase to elongate the strand with a truncated primer site. The complexity of these master mixes could pose disadvantages in trying to optimize reactions for crude samples of nucleic acids with numerous potential inhibitors. However, their speed and sensitivity gives them an advantage. They also produce short amplicons that could be used in detection approaches other than fluorescence dyes or probes.

Strand Displacement of Circular/Circularized Targets. The basic rolling circle amplification reaction is the production of a linear single strand amplicon from circular or circularized DNA using a single primer and a strand displacing polymerase. Various detection technologies and its low reaction temperature make this technique attractive for practitioner detection assays. However, the requirement for a circular template often means increased protocol complexity due the need for additional primers, enzymes, and associated processing steps. This method could be particularly useful when targets are on bacterial plasmids or some viruses or viroids.

Strand Displacement with Multiple Linear Primer Sets. This group of assays uses multiple primers sets to the same target region of a linear target DNA. These methods [e.g. Loop mediated isothermal amplification (LAMP), cross-priming amplification (CPA), and Smart amplification (Smart-AMP)] have been extensively investigated for the use in the detection of numerous plant pathogens (Tomlinson 2010). They have numerous advantages that will likely make them suitable for practitioner performed detection assays.

LAMP amplifies DNA with high specificity, sensitivity, and rapidity without thermal denaturation (Nagamine et al. 2001) under isothermal conditions (60–65 °C) thus only requiring inexpensive heat sources (e.g. water bath or block heater) (Notomi et al. 2000). Another advantage is that a large amount of magnesium pyrophosphate is precipitated as a byproduct of the reaction that enables visual assessment of amplification in real time or at the end of the reaction (Mori et al. 2001; Mori et al. 2004). There is also the potential of using fluorescent dyes and black light to further enhance the visual assessment of a positive reaction (Ohtsuka et al. 2005; Dukes et al. 2006; Mori et al. 2006). LAMP has been shown to be less sensitive to PCR inhibitors, thus requiring less DNA purification (Poon et al. 2006). The amount of product generated is relatively independent of initial target concentration, thus allowing for unambiguous determination of positive or negative reactions (Tomlinson et al. 2007). LAMP assays have the potential to be quantitative when used in conjunction with relatively inexpensive equipment (Mori et al. 2004; Tomlinson et al. 2007; Jenkins et al. 2011; Kubota et al. 2011a, b). LAMP has also been combined with a generic lateral flow device (Tomlinson 2010) to further simplify detection and increase the potential for practitioner-preformed LAMP assays.

Case History: Monitoring Grape Powdery Mildew

In the late 1990s using trap plants, we demonstrated that airborne inoculum of *E. necator* (grape powdery mildew) did not appear to be present in the air within vineyards in the Willamette Valley of Oregon, USA until well after bud break despite the environmental conditions being suitable for ascospore release (Pscheidt et al. 2000). Grove (2004) also demonstrated that ascospore release in Eastern Washington was largely governed by moisture and temperature. These data indicated that there was an asynchrony in host and pathogen development that could be exploited to reduce fungicide inputs for the management of grape powdery mildew. Over the course of several years, Falacy et al. (2007) demonstrated that qualitative PCR could be used to detect presence of airborne inoculum of *E. necator* in the beginning of the growing season and that this inoculum was associated with discrete wetness events (>2.5 mm of rainfall or overhead irrigation and average daily temperature >5 °C). However, in the Willamette Valley the pattern of ascospore release was very different (Hall 2000). Release often occurred during the winter when the host was dormant or well after bud break (Pscheidt et al. 2000; Thiessen et al. 2014). These data indicated that monitoring airborne inoculum could be used to initiate a fungicide program and potentially reduce the number of fungicide applications made over a growing season.

In 2005, we began exploring whether inoculum detection for initiating fungicide applications could be commercially implemented. Due to practitioner reluctance in risking a high value crop (>$20,000/ha), 6–11 rotating-arm impaction traps were initially placed within or around various vineyards. One trap was always placed within the region of the vineyard that had the most severe disease the previous season. Prior to 15 cm of shoot growth the traps were placed adjacent to grapevine trunks because the cleistothecia overwinter on grape bark (Pearson and Gadoury 1987). Traps were sampled every Monday and Thursday and processed for qualitative PCR (prior to 2008) or quantitative PCR (after 2008) the same day as collected. This sampling protocol resulted in two samples per generation time [assuming optimal conditions (Delp 1954)] which increased practitioner confidence of being able to manage an epidemic if the initial ascospore release was not detected. The practitioners were also provided with weekly disease scouting results. For the first 2 years, the practitioners were only asked to examine the data provided to them. Beginning in 2007, some vineyard managers agreed to delay fungicide applications to a few rows within vineyard blocks until airborne *E. necator* inoculum was detected. Over the course of 5 years, the vineyard area that was managed based on inoculum detection increased for each grower until the experimental unit was an individual management unit (generally 1–3 ha blocks); with one grower managing 40 ha from a single trap placed in the perennial "hotspot" (area of the vineyard that always has the first signs and usually the highest disease levels) over a 3 year period. On average 2.3, fungicide applications/year were eliminated across the 43 commercial locations and 5 years of testing. These results indicated that using inoculum detection was a viable management practice. There was also considerable practitioner interest in using the technology. However, its implementation was impeded by the cost of collecting and processing the trap samples in regions with lower concentration of vineyards. For example, Coastal Viticulture Consultants, Inc. was able to successfully market their services in monitoring *E. necator* inoculum using our trap design and quantitative PCR methods in the highly concentrated vineyard production of the Napa Valley of California but not in the sparse production area of the Willamette Valley of Oregon.

In 2008, we began development of a LAMP assay (Thiessen et al. 2014) to monitor the presence of *E. necator* inoculum. The assay proved to be extremely sensitive and robust when conducted using reagents and materials under proper conditions in the lab but was also suitable for practitioner use. To test the ability of practitioners to perform the LAMP assay, three spore traps were placed in the vineyard. One trap was completely maintained by the practitioner and used to perform the DNA extraction and LAMP assay at their facility and two traps were maintained by the lab where one was processed for quantitative PCR and the other was processed using the LAMP protocol employed by the practitioner. For these analyses, we assumed that the quantitative PCR was accurate (e.g. "the gold standard"). The lab performed LAMP was accurate 89 % of time compared to the quantitative PCR results with a sensitivity of 78 %, indicating that false negatives were more common than false positives. However, implementation by growers in 2011 was not as successful as in 2010 with an accuracy of 66 % compared to 74 % in 2010. Some of the discrepancy between lab and grower assays may have been due to differences in actual spore numbers collected on the qPCR and LAMP traps since many of the disagreements occurred at or near detection limits (<10 spores). The most likely cause of the difference in performance between the growers and qPCR is that reagent storage was less than optimal (e.g. frost free freezers or small freezers where reagents were not stored at a constant –20 °C) at most practitioner facilities. This difference likely resulted in degradation of the reagents due to freeze/thaw cycles during storage. These issues have been addressed by storing all reagents in insulated cryoboxes to reduced temperature fluctuations that occur in various freezers. There is still the possibility that some of the error was due to the growers being nervous about missing positive reactions and seeing subtle changes in turbidity when there really weren't any. These results indicated that the proposed LAMP assay was not appropriate for implementation using turbidity to determine reaction results. For these reasons, we pursued the development of a more robust detection system.

We are now collaborating with Daniel Jenkins at the University of Hawaii at Manoa and Ryo Kubota (Diagenetix, LLC) to pursue the development of a quantitative LAMP assay (Jenkins et al. 2011; Kubota et al. 2011a, b) using their Gene-DART and Smart-DART technology. This system is based on an assimilating probe that is specific for the loop portion of the *E. necator* LAMP amplicon and fluoresces upon amplification. The Gene-DART is a two part fluorescent probe master mix consisting of the assimilating probe, LAMP primers, and the Optigene Isothermal Master Mix, no dye (Optigene, Ltd. West Sussex, United Kingdom). In our experience, the Optigene master mix is stable at room temperature for several weeks, making it highly suitable for field applications. Smart-DART is the hand-held device that runs the LAMP reaction and records fluorescence over time using an android based smart phone or tablet computer. We have already developed and tested probes for both the forward and reverse loop of the *E. necator* LAMP amplicon. Using this approach, we were able to detect one conidium with no ambiguity in less than 30 min. We will be testing the Smart-DART and Gene-DART technologies with vineyard managers in the 2014 growing season.

Future

Our ability to detect and quantify airborne inoculum is rapidly improving and becoming more accessible to practitioners or automation at the field level. As this technology evolves, we will see an increased reliance on inoculum monitoring to inform management decisions. The use of inoculum monitoring will lead to increased precision in the application and timing of pesticides and reduce the environmental and economic costs associated with disease management and crop losses. In order to fully capitalize on the promise of this technology, it will require continued cross disciplinary research that incorporates research teams of nanotechnologists, physicists, engineers, computer scientists, mathematicians, with the more traditional collaborations among the various biological disciplines.

The future will likely include the real-time quantification of inoculum and disease presence in crops using a combination of the molecular identification, olfactory sensing and/or spectral imagery as farm equipment, probably autonomous, moves through the field. These data will then be integrated in real time with modeling systems for crop growth, pathogen dispersion and disease development to generate spatially explicit disease risk that will then be used to guide precision applications of cultural and pesticide management practices.

Acknowledgments This work was supported by USDA-ARS CRIS 5358-22000-039-OOD, American Vineyard Foundation, Oregon Wine Board, and Viticulture Consortium -West. I thank Tyrone Hall, David Gent, Gary Grove, Ken Johnson, Jessica Keune, Tara Neill, Jay Pscheidt, Todd Temple and Lindsey Thiessen for discussions that guided the ideas presented above. I also thank numerous practitioners that have helped refine the *E. necator* inoculum detection project and help bring it to commercialization. The use of trade, firm, or corporation names in this publication is for the information and convenience of the reader. Such use does not constitute an official endorsement or approval by the United States Department of Agriculture or the Agricultural Research Service of any product or service to the exclusion of others that may be suitable.

References

Alvarez AJ, Buttner MP, Stetzenbach LD (1995) PCR for bioaerosol monitoring: sensitivity and environmental interference. Appl Environ Microbiol 61:3639–3644

Aylor DE (1989a) Temporal and spatial development of bean rust epidermis initiated from an inoculated line source. Phytopathology 78:146–151

Aylor DE (1989b) Aerial spore dispersal close to a focus of disease. Agr Forest Meteorol 47:109–122

Aylor DE (2003) Spread of plant disease on a continental scale: role of aerial dispersal of pathogens. Ecology 84:1989–1997

Berger RD (1969) A celery blight spray program based on disease forecasting. Fla State Horticulture Soc 42:107–111

Calderon C, Ward E, Freeman J, Mccartney A (2002) Detection of airborne fungal spores sampled by rotating-arm and Hirst-type spore traps using polymerase chain reaction assays. J Aerosol Sci 33:283–296

Campbell CL, Madden LV (1990) Monitoring epidemics: pathogen. APS Press, St. Paul, pp 75–105
Carisse O, Tremblay DM, Lévesque CA, Gindro K, Ward P, Houde A (2009) Development of a TaqMan real-time PCR assay for quantification of airborne conidia of Botrytis squamosa and management of botrytis leaf blight of onion. Phytopathology 99:1273–1280
Carisse O, Levasseur A, Van der Heyden H (2012) A new risk indicator for botrytis leaf blight of onion caused by Botrytis squamosa based on infection efficiency of airborne inoculum. Plant Pathol 61(6):1154–1164
Cortesi P, Ottaviani MP, Milgroom MG (2004) Spatial and genetic analysis of a flag shoot subpopulation of Erysiphe necator in Italy. Phytopathology 94:544–550
Cortesi P, Mazzoleni A, Pizzatti C, Milgroom MG (2005) Genetic similarity of flag shoot and ascospore subpopulations of Erysiphe necator in Italy †. Society 71:7788–7791
Craw P, Balachandran W (2012) Isothermal nucleic acid amplification technologies for point-of-care diagnostics: a critical review. Lab Chip 12:2469–2486
De Boer SH, López MM (2012) New grower-friendly methods for plant pathogen monitoring. Annu Rev Phytopathol 50:197–218
Delp CJ (1954) Effect of temperature and humidity on the grape powdery mildew fungus. Phytopathology 44:615–626
Dhingra OD, Sinclair JB (1995) Basic plant pathology methods. CRC Press, Boca Raton
Dukes JP, King DP, Alexandersen S (2006) Novel reverse transcription loop-mediated isothermal amplification for rapid detection of foot-and-mouth disease virus. Arch Virol 151:1093–1106
Falacy JS, Grove GG, Mahaffee WF, Galloway H, Glawe DA, Larsen RC, Vandemark GJ (2007) Detection of Erysiphe necator in air samples using the polymerase chain reaction and species-specific primers. Phytopathology 97:1290–1297
Fröhlich-Nowoisky J, Pickersgill DA, Després VR, Pöschl U (2009) High diversity of fungi in air particulate matter. Proc Natl Acad Sci U S A 106:12814–12819
Gent DH, Turechek WW, Mahaffee WF (2007) Sequential sampling for estimation and classification of the incidence of hop powdery mildew I: leaf sampling. Plant Dis 91:1002–1012
Gent DH, Nelson ME, Farnsworth JL, Grove GG (2009) PCR detection of Pseudoperonospora humuli in air samples from hop yards. Plant Pathol 58:1081–1091
Gill P, Ghaemi A (2008) Nucleic acid isothermal amplification technologies: a review. Nucleosides Nucleotides Nucleic Acids 27:224–243
Gregory PHE (1973a) The microbiology of the atmosphere. Wiley, New York
Gregory PHE (1973b) Outdoor aerobiology. Oxford University Press, London, p 16
Grove GG (2004) Perennation of Uncinula necator in Vineyards of Eastern Washington. Plant Dis 88:242–247
Gubler WD, Rademacher MR, Vasquez SJ, Thomas CS (1999) Control of powdery mildew using the uc davis powdery mildew mask index. Available from http://www.apsnet.org/online/feature/pmildew/Top.html
Hall TW (2000) Epidemiology of grape powdery mildew, Uncinula necator, in the Willamette Valley. Thesis, Oregon State University, Corvallis, OR
Hardwick NV (2006) In: Cooke BM, Jones DG, Kaye B (eds) Disease forecasting: the epidemiology of plant diseases. Springer, Dordrecht, pp 239–267
Harrison MD, Livingston CH, Oshima N (1965) Control of potato early blight. II. spore traps as a guide for initiating applications of fungicides. Am Potato J 42:333–340
Hershman DE, Sikora EJ, Giesler LJ (2011) Soybean rust PIPE: past, present, and future. J Integr Pest Manage 2:1–7
Jacobi C, Backman PA, Davis DP, Fellow FP (1995) AU-Pnuts advisory 1: development of a rule-based system for scheduling peanut leaf spot fungicide applications. Plant Dis 79:666–671
Jacobsen BJ (2007) Ecologically based management of plant diseases. In: Cuperus OKG (ed) Ecologically based integrated pest management. CAB International, New York, p 462
Jaenicke R (2005) Abundance of cellular material and proteins in the atmosphere. Science (New York, N.Y.) 308:73

Jenkins DM, Kubota R, Dong J, Li Y, Higashiguchi D (2011) Handheld device for real-time, quantitative, LAMP-based detection of Salmonella enterica using assimilating probes. Biosens Bioelectron 30:255–260

Kast W (2000) Experience about the OiDiag-system against Powdery Mildew (*Uncinula necator*) in 1994–1997. In: Magarey PA, Thiele SA, Tschirpig KL, Emmett RW, Clarke K, Magarey RD (eds) Third International Workshop on grapevine downy and powdery mildew, SARDI research report series no. 22. SARDI, Adelaide, pp 619–626

Kennedy R, Wakeham A, Byrne KG, Meyer UM, Dewey FM (2000) A new method to monitor airborne inoculum of the fungal plant pathogens *Mycosphaerella brassicicola* and *Botrytis cinerea*. Appl Environ Microbiol 66:2996–3000

Khattab A, Levetin E (2008) Effect of sampling height on the concentration of airborne fungal spores. Ann Allergy Asthma Immunol 101:529–534

Kremheller HT, Diercks R (1983) Epidemiologie und Prognose des Falschen Mehltaues (Pseudopernospora humuli) an Hopfen. Zeitschrift Für Pflanzenkrankheiten Und Pflanzenschutz 90:599–616

Kubota R, Alvarez AM, Su WW, Jenkins DM (2011a) FRET-based assimilating probe for sequence-specific real-time monitoring of loop-mediated isothermal amplification (LAMP). Biol Eng Trans 4:81–100

Kubota R, Labarre P, Singleton J, Beddoe A, Weigl BH, Alvarez AM, Jenkins DM (2011b) Non-instrumented nucleic acid amplification (NINA) for rapid detection of Ralstonia solanacearum race 3 biovar 2. Biol Eng Trans 4:69–80

Kuske CR, Banton KL, Adorada DL, Stark PC, Hill KK, Jackson PJ (1998) Small-scale DNA sample preparation method for field PCR detection of microbial cells and spores in soil. Society 64:2463–2472

Lacey ME, West JS (2006) The air spora: a manual for catching and identifying airborne biological particles. Springer, Dordrecht

Lee S-A, Adhikari A, Grinshpun SA, McKay R, Shukla R, Reponen T (2006) Personal exposure to airborne dust and microorganisms in agricultural environments. J Occup Environ Hyg 3:118–130

Madden LV, Hughes G (1999) Sampling for plant disease Incidence. Phytopathology 89:1088–1103

Madden L, Hughes G, van den Bosch F (2007) The study of plant disease epidemiology. APS Press, St. Paul

Magarey RD, Travis JW, Russo JM, Seem RC, Magarey PA (2002) Decision support systems: quenching the thirst. Plant Dis 86:4–14

McCartney H, Fitt B, Schmechel D (1997) Sampling bioaerosols in plant pathology. J Aerosol Sci 28:349–364

McDevitt JJ, Lees PSJ, Merz WG, Schwab KJ (2004) Development of a method to detect and quantify Aspergillus fumigatus conidia by quantitative PCR for environmental air samples. Mycopathologia 158:325–335

Mori Y, Nagamine K, Tomita N, Notomi T (2001) Detection of loop-mediated isothermal amplification reaction by turbidity derived from magnesium pyrophosphate formation. Biochem Biophys Res Commun 289:150–154

Mori Y, Kitao M, Tomita N, Notomi T (2004) Real-time turbidimetry of LAMP reaction for quantifying template DNA. J Biochem Biophys Methods 59:145–157

Mori Y, Hirano T, Notomi T (2006) Sequence specific visual detection of LAMP reactions by addition of cationic polymers. BMC Biotechnol 6:3

Mundt CC (2009) Importance of autoinfection to the epidemiology of polycyclic foliar disease. Phytopathology 99:1116–1120

Nagamine K, Watanabe K, Ohtsuka K, Hase T, Notomi T (2001) Loop-mediated isothermal amplification reaction using a nondenatured template. Clin Chem 47:1742–1743

Nagarajan S, Singh DV (1990) Long-distance dispersion of rust pathogens. Annu Rev Phytopathol 28:139–153

Niemz A, Ferguson TM, Boyle DS (2011) Point-of-care nucleic acid testing for infectious diseases. Trends Biotechnol 29:240–250

Notomi T, Okayama H, Masubuchi H, Yonekawa T, Watanabe K, Amino N, Hase T (2000) Loop-mediated isothermal amplification of DNA. Nucleic Acids Res 28:E63

Ohtsuka K, Yanagawa K, Takatori K, Hara-kudo Y (2005) Detection of *Salmonella enterica* in naturally contaminated liquid eggs by loop-mediated isothermal amplification, and characterization of *Salmonella Isolates*. Society 71:6730–6735

Ojiambo P, Holmes G (2011) Cucurbit downy mildew ipmPIPE: a next generation web-based interactive tool for disease management and extension outreach. Plant Health Progress. doi: 10.1094/PHP-2011-0411-01-RV

Pearson RC, Gadoury DM (1987) Cleistothecia, the source of primary inoculum for grape powdery mildew in New York. Phytopathology 77:1509–1514

Poon LLM, Wong BWY, Ma EHT et al (2006) Sensitive and inexpensive molecular test for falciparum malaria: detecting *Plasmodium falciparum* DNA directly from heat-treated blood by loop-mediated isothermal amplification. Clin Chem 52:303

Posthuma-Trumple GA, Korf J, Van Amerongen A (2009) Lateral flow (inmuno)assay: its strengths, weakness, opportunities and threats. A literature survey. Anal Bioanal Chem 393:569–582

Pscheidt JW, Wittig HPP, Wallace LD, Hall T, Mahaffee WF (2000) In: Magarey Thiele SA, Tschirpig KL, Emmett RW, Clarke K, Magarey RD, P A, Evaluation of three grape powdery mildew forecasting programs in Corvallis, Oregon, USA, SARDI Research Report Series Adelaide, Australia pp 143–146

Rogers SL, Atkins SD, West JS (2009) Detection and quantification of airborne inoculum of Sclerotinia sclerotiorum using quantitative PCR. Plant Pathol 58:324–331

Rossi V, Caffi T, Legler SE, Cattolica U (2010) Dynamics of ascospore maturation and discharge in *Erysiphe necator*, the causal agent of grape powdery mildew. Phytopathology 100:1321–1329

Royle DJ (1973) Quantitative relationships between infection by the hop downy mildew pathogen, Pseudoperonospora humuli, and weather and inoculum factors. Ann Appl Biol 73:19–30

Sackett KE, Mundt CC (2005) The effect of dispersal gradient and pathogen life cycle components on epidemic velocity in computer simulations. Phytopathology 95:9

Schaad NW, Frederick RD, Shaw J, Schneider WL, Hickson R, Petrillo MD, Luster DG (2003) Advances in molecular-based diagnostics in meeting crop biosecurity and phytosanitary issues. Annu Rev Phytopathol 41:305–324

Schmale DG, Da S, Bergstrom GC (2005) Spatial patterns of viable spore deposition of Gibberella zeae in wheat fields. Phytopathology 95:472–479

Schneider RW, Haudenshield JS, Hartman GL, Mahaffee WF (2009) Development and use of fluorescent antibody and qPCR protocols for the electrostatic spore trap. Phytopathology 99: S115

Thiessen LD, Keune JA, Neill TM, Turechek WW, Grove GG, Mahaffee WF (2014) The development of a grower performed inoculum detection assay for management of grape powdery mildew *(Erysiphe necator)*. Plant Disease (submitted)

Tomlinson J (2010) Potential of LAMP for detection of plant pathogens. CAB Rev: Perspectives in Agriculture, Veterinary Science, Nutrition and Natural Resources 3:1–7

Tomlinson JA, Barker I, Boonham N (2007) Faster, simpler, more-specific methods for improved molecular detection of *Phytophthora ramorum* in the field. Appl Environ Microbiol 73:4040–4047

Tomlinson JA, Dickinson MJ, Boonham N (2010) Rapid detection of *Phytophthora ramorum* and *P. kernoviae* by two-minute DNA extraction followed by isothermal amplification and amplicon detection by generic lateral flow device. Phytopathology 100:143–149

Tong Y, Lighthart B (2000) The annual bacterial particle concentration and size distribution in the ambient atmosphere in a rural area of the Willamette Valley, Oregon. Aerosol Sci Technol 32:37–41

Turechek WW, Mahaffee WF (2004) Spatial pattern analysis of hop powdery mildew in the Pacific Northwest: implications for sampling. Phytopathology 94:1116–1128

Van der Heyden H, Lefebvre M, Roberge L, Brodeur L, Carisse O (2013) Spatial pattern of strawberry Powdery Mildew (Podosphaera aphanis) and airborne inoculum. Plant Dis 97 (in press)

West JS, Atkins SD, Emberlin J, Fitt BDL (2008) PCR to predict risk of airborne disease. Trends Microbiol 16:380–387

West J, Atkins S, Fitt B (2009) Detection of airborne plant pathogens; halting epidemics before they start. Outlooks Pest Manage 20(1):11–14

Proximal Sensing of Plant Diseases

Erich-Christian Oerke, Anne-Katrin Mahlein, and Ulrike Steiner

Abstract Proximal sensing techniques have a large potential in surveying crops for the occurrence of diseases varying in spatial and temporal distribution within crops. Incidence of plant diseases results from crop status, the presence of inoculum, and suitable abiotic environmental factors, and often is heterogeneous in the field. Various technical sensors may be suitable for the detection, identification and quantification of plant diseases on different scales. Thermography, fluorescence and spectral sensors are very promising, but other techniques like electronic nose may be also useful. The full potential of these advanced detector technologies may be exploited only in combination with innovative methods of data processing for the extraction of relevant information. These technologies may support further Integrated Pest Management programs for sustainable crop production.

Keywords Proximal sensing • Disease symptoms • Thermography • Fluorescence imaging • Spectral imaging • Image processing

Introduction

The worldwide demand for agricultural products currently exceeds the supply. In a world of limited resources agricultural production has to be managed more efficiently (Von Witzke et al. 2008). Innovative technologies for Precision agriculture may help to use the right input factors at the right site at the right time (Oerke et al. 2010). The production potential of crops is often reduced by the incidence of various pests which are responsible for considerable reductions in the quantity and quality of crop production – even in the presence of activities to control the pests (Oerke 2006).

E.-C. Oerke (✉) • A.-K. Mahlein • U. Steiner
INRES – Phytomedicine, University of Bonn, Nussallee 9, 53115 Bonn, Germany
e-mail: ec-oerke@uni-bonn.de

M.L. Gullino and P.J.M. Bonants (eds.), *Detection and Diagnostics of Plant Pathogens*, Plant Pathology in the 21st Century 5, DOI 10.1007/978-94-017-9020-8_4,

The incidence of pests impairing plant growth in the field is often rather heterogeneous in time and space – at least in early stages (Waggoner and Aylor 2000). In conventional farming, spatial heterogeneity of disease symptoms complicates the decision whether diseases have to be controlled. Crops have to be surveyed several times per growth period and often complex sampling schemes are needed to give representative results for the decision to apply fungicides uniformly in the field. Within-field heterogeneity of disease implies that plants at some sites may need disease control while plants at other sites do not. Innovative technologies now make it possible to extend sampling/surveying from a few (representative) plants to all plants of a field in order to spot primary disease patches (West et al. 2003). Automatable detection, identification and quantification of diseases on the plant scale are prerequisite for a site-specific application of fungicides, adequate to disease incidence and precise in space and time. Based on information from sensors and the use of geographic information systems (GIS), application maps can be generated or sensors are linked to actuators in even more sophisticated online systems. Site-specific application of pesticides has the potential to reduce pesticide use, and thus may cut the economical expenses and ecological impacts in agricultural crop production systems (Gebbers and Adamchuk 2010).

For sensory detection of plant diseases caused by microorganisms the sensitivity of the sensing technique and the specificity of the signal are crucial. The differentiation among diseases occurring on a crop is an essential feature for operational systems, and host plant colonization by pathogens may be detected – better identified – even before typical disease symptoms appear. Although stress causing agents, and likewise various pathogens often cause similar (disease) symptoms and changes in plant physiology (Nutter et al. 2010; Stafford 2000), more specific effects may be used for the identification of diseases. Technical sensors and the acquired information, respectively, therefore, need to have diagnostic properties.

Proximal Sensing *vs.* Remote Sensing

The sensitivity of sensors for small disease symptoms and the specificity of the signal strongly depend on spatial resolution. Early detection of low disease levels and diagnosis of the disease, therefore, require high spatial resolution currently available only from proximal sensing systems. Proximal sensing, in contrast to remote sensing, is used here for the non-destructive near-range assessment of plants without the need for substantial pre-processing – like geometric and atmosphere corrections – of data. Proximal sensors may be hand-held, machine-mounted or attached to suitable unmanned aerial vehicles (UAVs), whereas satellites, aircrafts and UAVs covering larger areas are typical remote sensing platforms (Fig. 1). Spatial resolution in proximal sensing typically is in the range from 10^{-3} to 10^{-2} m, in contrast to 10^{-1} to 10^{2} m for sensors from aircrafts and satellites used in remote sensing.

Remote sensing refers to the assessment of (spectral) information from an object without making physical contact and has been the pioneer in most sensing

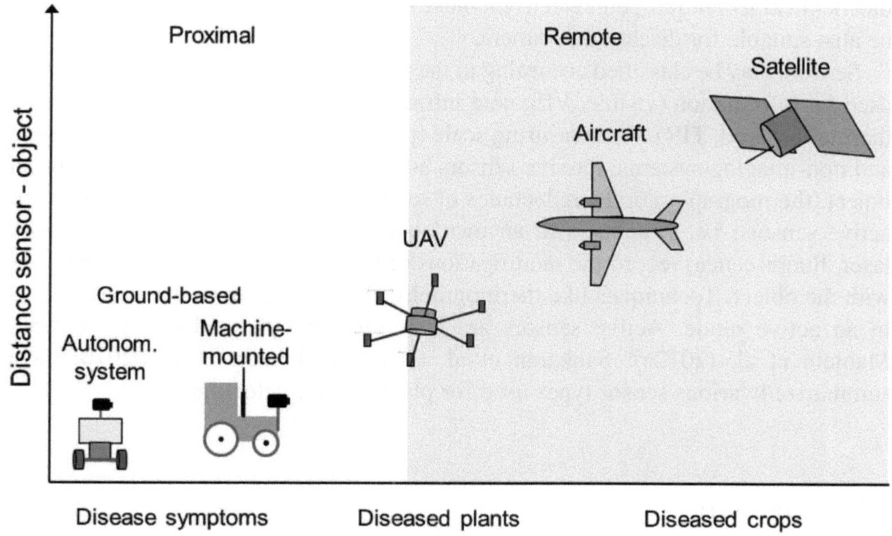

Fig. 1 Effect of sensor platform and spatial resolution on the potential of sensors to detect and identify plant diseases

technologies currently available. Remote sensing of diseased crops from space-borne and air-borne platform has been shown to be successful in later stages of disease epidemics (Franke and Menz 2007; Nutter et al. 2010; Huang et al. 2012), and need additional ground-truth data for signal interpretation. Airborne thermography has been successfully used for the delineation of management zones within fields – based on the transpiration of crop canopies related to plant biomass – as well as for the assessment of plant vigor related to crop yield in fungicide trails (Lenthe 2006; Stenzel et al. 2007).

Air-borne sensors are highly suitable for studying the spatial patterns of diseases on larger scales as well as over time. The use of airborne sensors for crop protection in the field, however, is currently limited by spatial resolution and temporal availability of (sensor platforms and) data.

Systems for Proximal Disease Sensing

Non-invasive technical sensor systems can provide detailed and highly resolved information on crops and individual plants. Several sensor types have been tested for their suitability to detect early changes in plant physiology due to biotic and abiotic stresses (Mahlein et al. 2012a; Sankaran et al. 2010). Assessment of the interactions of electromagnetic radiation with plant tissue currently includes the most promising techniques, especially measurements of plant reflectance, fluorescence and thermography. However, other techniques ranging from mechanical to biochemical techniques, e.g. electronic nose (e-nose) and headspace analytics may be also suitable for disease assessment.

Sensors may be classified according to the range in the electromagnetic spectrum used for information (visible, VIS; near infrared, NIR; short wave infrared, SWIR; thermal infrared, TIR), the measuring scale (proximal or remote), and into imaging and non-imaging systems. Passive sensors assess the characteristic radiation of an object (thermography) or the reflectance of solar radiation, whereas the detectors of active sensors, i.e. systems with an own source of radiation (reflectance, radar, laser, fluorescence) record the modifications of the radiation due to its interactions with the object. Techniques like thermography may be used in a passive as well as in an active mode. Active sensors are less affected by environmental factors. Mahlein et al. (2012a); Sankaran et al. (2010), and West et al. (2010) have summarized various sensor types used for plant disease detection.

Thermography

Thermal imaging is a non-contact technique to determine the temperature distribution of any object in a short period of time (Vadivambal and Jayas 2011). Infrared radiation emitted from plant surfaces may be recorded by detectors sensitive to radiation in the TIR from 8 to 12 μm. Each pixel of the images is related to a temperature value of the object's surface and may be illustrated in false color images. The performance of thermographic cameras is characterised by thermal sensitivity (thermal range, measurement precision), image resolution (image pixel size, number of image pixels, depth of focus), and scan speed (Oerke and Steiner 2010). The technology can be used from microscope applications to ground-based equipment covering a range from (leaf) tissue to crop canopies.

The temperature of plant tissue is determined by temperature of the environment and its water status regulating the transpiration rate; leaf temperature increases as transpiration rate decreases (Chaerle and van der Straeten 2000; Jones and Schofield 2008; Oerke and Steiner 2010). Similar to abiotic factors, pathogens may affect the status of stomata regulating the temperature gradient between plant tissue and air temperature. Spatial and temporal patterns of within-tissue heterogeneities in transpiration resulting from plant diseases can be imaged, sometimes even before typical disease symptoms appear, and may be monitored during disease spread.

Localised temperature changes due to pathogen attack or defense mechanism of the host plant have been reported for tobacco infected with tobacco mosaic virus (Chaerle et al. 2007), sugar beet infected by *Cercospora beticola* (Chaerle et al. 2007), downy mildew of cucumber caused by *Pseudoperonospora cubensis* (Lindenthal et al. 2005; Oerke et al. 2006), grapevine leaves infected with *Plasmopara viticola* (Stoll et al. 2008), and for apple leaves infected by *Venturia inaequalis* (Oerke et al. 2011).

On the field scale thermography may be used for the detection of primary disease foci and for the measurement of environmental conditions like leaf wetness

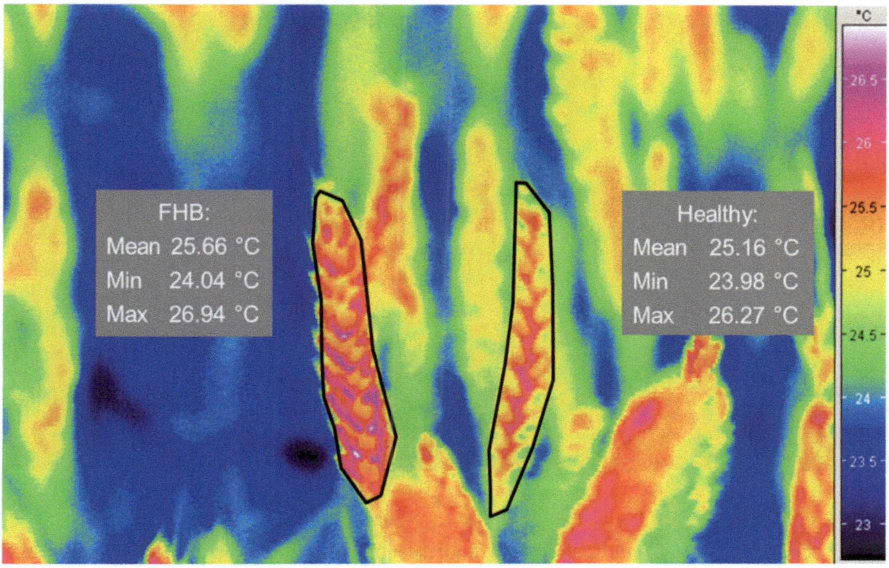

Fig. 2 Thermographic detection of *Fusarium* head blight-infected wheat ears at growth stage 75 in the field; comparison of temperature of infected and non-infected ears

duration which may be used in models calculating disease risk (Lindenthal 2005; Lenthe et al. 2007). The incidence of monocyclic diseases like *Fusarium* head blight could be spotted in a wheat canopy as infected ears showed premature ripening compared to uninfected ears (Fig. 2; Oerke and Steiner 2010). Thermography is a useful tool to monitor the incidence of soil-borne pathogens which often affect the water status of the shoot. Several nematodes and soil-borne virus diseases are characterized by low mobility and patchiness of above-ground symptoms in the field (Hillnhuetter et al. 2011a; Schmitz et al. 2004).

Thermography is highly sensitive to environmental conditions during measurements and the potential of thermography for precision disease control is limited. Despite of its high sensitivity, the thermal response (=change in transpiration) of plants to pathogen attack largely lacks diagnostic potential for the identification of diseases.

Fluorescence Measurements

Various fluorescence parameters of plants irradiated with ambient excitation light may be recorded for the assessment of photosynthetic activity and the content of chlorophyll and other plant metabolites, e.g. phenols. These methods are very sensitive to detect changes in photosynthesis (Fig. 3; Scholes and Rolfe 2009; Csefalvay et al. 2009). Since disease development also affects the crops

Fig. 3 Color image (*left*) and chlorophyll fluorescence parameter q_N (*right*) of an apple leaf infected by *Venturia inaequalis*

photosynthetic apparatus – pigments, electron transport chain, enzymes of the CO_2 fixing Calvin cycle – the intensity as well as the spectrum of chlorophyll fluorescence are modified in diseased plants, sometimes even before visible symptoms appear (Chaerle et al. 2009; Kuckenberg et al. 2009; Buerling et al. 2011). Recently the potential of chlorophyll fluorescence image analysis for quantitative phenotyping of plant resistance has been demonstrated (Rousseau et al. 2013). Modification in photosynthesis is a general response of plants to various stress factors; therefore, chlorophyll fluorescence also lacks diagnostic potential. Measurements of blue light-induced fluorescence and of fluorescence spectra may be more suitable for disease identification but information is still very low (Bellow et al. 2013; Buerling et al. 2012). The use of active sensors like these in the field is also limited by the response time, i.e. the time between emission and detection of the measuring beam, limiting the use for on-the-go sensing.

Spectral Techniques

Reflectance of incoming electromagnetic radiation in the visible (VIS, 400–700 nm), near infrared (NIR, 700–1,100 nm) and short wave infrared (SWIR, 1,100–2,500 nm) depends on multiple interactions: back scattering at the leaf surface and internal cellular structures, radiant energy absorption induced by leaf chemistry, e.g. content of pigments, leaf water, proteins or carbon (Jacquemoud and Ustin 2001).

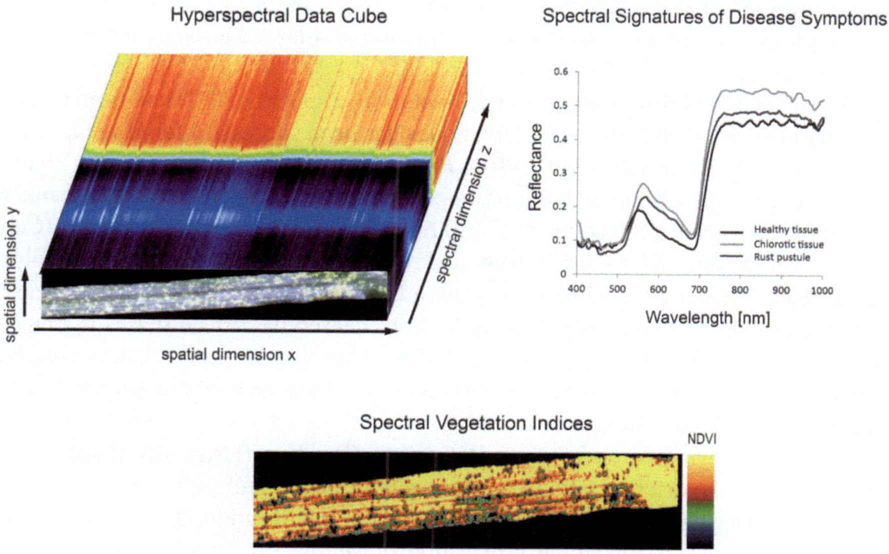

Fig. 4 Hyperspectral imaging of barley leaf diseased with brown rust. Each pixel of the 3-dimensional data cube contains a continuous spectral signature from 400 to 1,000 nm. Based on this imaging data spectral vegetation indices maps may be calculated

The evolution of spectral sensors started with multispectral sensors – with <10 discrete wavebands which do not produce the spectrum of an object – to hyperspectral sensors – with many narrow spectral bands (<10 nm) over a continuous spectral range producing a spectrum – and ultraspectral sensors with even more wavebands. With spectral resolution sometimes >1 nm these sensors provide a multiplicity of information from 350 to 2,500 nm. Space-borne and aircraft-mounted remote systems have lower spatial resolution than proximal systems and are hardly suitable for the detection of disease symptoms on the leaf or plant scale. Spatial resolution, hence the smallest identifiable symptom or structure, is defined by the minimum size of a pixel and should be better than 1 mm for early disease symptoms. Non-imaging detectors merge spectral information from an area depending on the sensor's field of view to one spectrum; the mixture of information from disease symptoms, non-diseased plant tissue and background hardly allows conclusions on the health status or disease of the plant material (Mahlein et al. 2010; Steiner et al. 2008). In contrast, spectral imaging systems provide spectral information (z-axis) for all pixels of the image (axes x and y). Each pixel contains the full spectral information ranging from UV to NIR and SWIR, depending on the detector (Fig. 4).

Diseases result in modified optical properties of leaf tissue due to changes in the structure and pigmentation of plant tissue, and pathogen structures produced on and within the leaf surface. Disease-specific symptoms, e.g. spatial and temporal patterns of chlorotic and necrotic tissue varying in optical characteristics as well as in chemical composition, and mycelium and spores of powdery mildew, rust, and downy mildew may be detectable. In later stages plant morphology, crop canopy morphology and density as well as the interaction of solar radiation with plants may be affected (West et al. 2010).

The detection of diseased plants, i.e. plants with a spectrum different from that of healthy ones, using spectroscopic techniques has been successful for *Magnaporthe grisea* on rice (Kobayashi et al. 2001), *Phytophthora infestans* on tomato (Zhang et al. 2003), *Venturia inaequalis* on apple trees (Delalieux et al. 2007), canker lesions on citrus fruits (Quin et al. 2009), *Blumeria graminis* on barley (Cao et al. 2013), and *Rhizoctonia* root and crown rot of sugar beet (Reynolds et al. 2012). In some experiments a relation between spectral changes and disease severity could be established. Damage to crops caused by virus diseases in grapevine (Naidu et al. 2009) or insects in maize (Carrol et al. 2008) could also be detected. Infections of sugar beet by various leaf pathogens could be detected even pre-symptomatically (Rumpf et al. 2010).

As specific symptoms vary from disease to disease, different wavebands may be suitable for the discrimination among diseases which is still a challenge. The VIS and NIR range gave the best information for disease identification. Bravo et al. (2003) used spectral images for an early detection of yellow rust in wheat in the field, Hillnhuetter et al. (2011b) successfully discriminated above-ground symptoms of sugar beet caused by *Heterodera schachtii* and *Rhizoctonia solani*. Spectral signatures of sugar beet leaves with *Cercospora* leaf spot, powdery mildew and sugar beet rest were significantly different and disease specific (Mahlein et al. 2010, 2013). Spectral imaging provided disease-specific signatures of leaves of sugar beet and barley infected by respective leaf pathogens (Mahlein et al. 2013; Mahlein et al. unpublished).

Non-optical Sensors

Plants are reported to release a range of volatile organic compounds (VOCs, Dudareva et al. 2006). The VOC profile depends on plant development and the abiotic (temperature, relative humidity, light, nutrient availability) and biotic (pathogens, animal pests) environment. Plant VOCs produced in response to pathogen attack may be used for the detection and identification of primary disease foci. Currently the use of gas chromatography and electronic noses are tested for their suitability under controlled conditions, e.g. post-harvest disease (Li et al. 2009; Prithiviraj et al. 2004; Kushalappa et al. 2002), on leaves (Laothawornkitkul et al. 2008) as well as in the field (Spinelli et al. 2006; Markom et al. 2009). Irrespective of disease sensitivity and disease specificity these techniques would be able to spot the site of diseased plants in the field by following the gradient of emitted VOCs.

Recently terahertz spectroscopy and NMR- and x-ray-based imaging technologies have been also used for the detection and differentiation of plant diseases (Pearson and Wicklow 2006; Hadjiloucas et al. 2009; Narvankar et al. 2009;

Hillnhuetter et al. 2012). Due to the complex equipment application is limited to research studies by now.

Processing and Analysis of Sensor Data

Powerful methods of data analysis are crucial in the use of optical sensors for the detection, differentiation and quantification of plant diseases. Recording of large amounts of information with high spectral, spatial and temporal resolution on the object is characteristic for imaging systems, especially for hyperspectral systems.

In **thermography**, the difference between leaf tissue and air temperature and the maximal temperature difference (MTD) within the plant tissue of interest are more suitable than the absolute temperature (Oerke et al. 2006; Oerke and Steiner 2010). Isotherms may be used to visualize hot and cold spots due to infections by leaf pathogens (Oerke et al. 2011). Boquete et al. (2010) applied independent component analysis on thermal infrared images for an automated detection of high tumor risk area in human cancer research and this approach may be adapted to plant pathology.

Effects of diseases on **spectral reflection** may be assessed and analysed using simple statistical methods or by more complex data mining and classification algorithms. E.g. differences between spectral signatures can be easily calculated as difference spectra, ratios or derivations (Carter and Knapp 2001; Pietrzykowski et al. 2006). Mahlein et al. (2010) identified significantly different regions in difference spectra from healthy sugar beet and plants diseased with *Cercospora* leaf spot, powdery mildew and rust.

Spectral vegetation indices (SVIs) are widely used for monitoring, analysing, and mapping of temporal and spatial variation in vegetation. These algorithms have been developed in remote sensing of vegetation and are based on specific wavelengths of the spectra related to biochemical and biophysical plant parameters indicating color, material, biomass, etc. (Blackburn 2007; Thenkabail et al. 2000). Characteristically SVIs result in a reduction of data dimension and may be used for the quantification of pigments and biomass. SVIs from multispectral data were useful for estimating the severity of *Cercospora* leaf spot in sugar beet fields and were in agreement with visual ratings from plant pathologists (Steddom et al. 2005). Delalieux et al. (2007) used SVIs for the assessment of apple scab at different stages of disease development. Combinations of two or more indices improved the discrimination between three diseases of sugar beet from non-imaging spectroscopy (Mahlein et al. 2010). SVIs developed for remote sensing are sensitive to changes in leaf reflectance resulting from plant diseases, however, they are hardly disease-specific and can be used for disease identification only by introducing disease-specific threshold values. The development of spectral disease indices with high sensitivity to spectral characteristics of diseases has shown their potential to improve the use of spectral indices for plant disease detection and identification (Mahlein et al. 2013).

In remote sensing applications, classification is the assignment of a spectral signature to a characteristic group or class, and is used to discriminate groups from each other. Various unsupervised and supervised classification techniques have been successfully used for the detection of changes in spectral reflectance associated with the development of disease symptoms. Spectral angle mapper (SAM), discriminant analysis, and machine learning methods, e.g. k-means clustering, artificial neural networks (ANN) and support vector machines (SVM) are commonly used for data analysis (Rumpf et al. 2010; Sankaran et al. 2010). The SAM, i.e. comparing spectral similarity by calculating the angular difference between a spectrum and a reference spectrum in a n-dimensional space has shown high overall accuracy for the classification of diseases in sugar beet and barley (Mahlein et al. 2012b, unpublished). The quality and interpretation of classification depends on the suitability of data representation. Commonly optical sensors acquire more information than required for classification, thus different feature selection algorithms or methods for feature reduction are applied prior to classification. One widely-used method is the principal component analysis (PCA) transforming high dimensional data to few principal components with descending variance (Bauriegel et al. 2011). Alternatively SVIs may be selected as features for an automatic classification of plant diseases with data mining algorithms as demonstrated by Rumpf et al. (2010) for early detection of sugar beet diseases. Data mining techniques are reported to be superior in multi-class approaches (Moshou et al. 2004; Rumpf et al. 2010).

Although all these algorithms have their own merits, there is not a single approach optimal for all applications; a direct comparison of analysis methods is unfeasible in many cases. Although classification of hyperspectral images commonly ignores information from adjacent pixels by now (Plaza et al. 2009), the multidimensional analysis of spectral and spatial patterns has a high potential in future, especially for the interpretation of highly-resolved images of disease symptoms. But large file sizes and complex algorithms also imply high computing times.

Fields of Application

Reliable, precise and accurate assessment of diseases is important for predicting yield loss, monitoring and forecasting epidemics and for understanding fundamental biological processes (Bock et al. 2010). Currently only spectral sensors (and VOC sensors) have diagnostic properties required for the differentiation/identification of plant diseases whereas other non-invasive proximal sensors with high sensitivity to metabolic changes may be used also for disease detection (Table 1). Optical sensor technologies may be implemented at various scales and for different purposes, e.g. resistance breeding in the field and under controlled conditions, screening of new fungicidal compounds, precision disease control in the field.

Table 1 Suitability of various sensor types for the detection, identification and quantification of plant diseases and their availability under field conditions

	Disease			
Sensor	Detection	Identification	Quantification	Availability in the field
Spectral	++(+)	+++	+++	++
Thermal	+++	+	++	++
Fluorescence	+++	+	++	+
Electronic nose	+++	++	+	(+)
Others[a]	++	(+)	+	–

[a]NMR, X-ray, etc

In large-scale experiments sensor techniques may contribute to speed up screening assays in resistance breeding (Montes et al. 2007). Inoculation of plants with a pathogen at a well-known time creates optimal conditions for an automatic sensor system with high sensitivity and reproducibility. Since disease identification is not required in inoculation experiments, also thermography and fluorescence techniques seem to be suitable for these applications. Spectral imaging microscopy may result in new spatial information on structural and chemical modifications of the plant tissue in the course of host-pathogen interactions.

References

Bauriegel E, Giebel A, Geyer M, Schmidt U, Herppich WB (2011) Early detection of *Fusarium* infection in wheat using hyperspectral imaging. Comput Electron Agric 75:304–312

Bellow S, Latouche G, Brown SC, Poutaraud A, Cerovic ZG (2013) Optical detection of downy mildew in grapevine leaves: daily kinetics of autofluorescence upon infection. J Exp Bot 64:333–341

Blackburn GA (2007) Hyperspectral remote sensing of plant pigments. J Exp Bot 58:844–867

Bock CH, Poole GH, Parker PE, Gottwald TR (2010) Plant disease severity estimated visually, by digital photography and image analysis, and by hyperspectral imaging. Crit Rev Plant Sci 29:59–107

Boquete L, Ortega S, Miguel-Jienez JM, Rodriguez- Ascariz JM, Blanco R (2010) Automated detection of breast cancer in thermal infrared images, based on independent component analysis. J Med Syst 36:103–111

Bravo C, Moushou D, West J, McCartney A, Ramon H (2003) Early disease detection in wheat fields using spectral reflectance. Biosyst Eng 84:137–145

Buerling K, Hunsche M, Noga G (2011) Use of blue-green and chlorophyll fluorescence measurements for differentiation between nitrogen deficiency and pathogen infection in wheat. J Plant Physiol 168:1641–1648

Buerling K, Hunsche M, Noga G (2012) Presymptomatic detection of powdery mildew infection in winter wheat cultivars by laser-induced fluorescence. Appl Spectrosc 66:1411–1419

Cao X, Luo Y, Zhou Y, Duan X, Cheng D (2013) Detection of powdery mildew in two winter wheat cultivars using canopy hyperspectral reflectance. Crop Prot 45:124–131

Carrol MW, Glaser JA, Hellmich RL, Hunt TE, Sappington TW, Calvin D et al (2008) Use of spectral vegetation indices derived from airborne hyperspectral imagery for detection of European corn borer infestation in Iowa corn plots. J Econ Entomol 101:1614–1623

Carter GA, Knapp AK (2001) Leaf optical properties in higher plants: linking spectral characteristics to stress and chlorophyll concentration. Am J Bot 88:677–684

Chaerle L, Van der Straeten D (2000) Imaging techniques and the early detection of plant stress. Trends Plant Sci 5:495–501

Chaerle L, Leinonen I, Jones HG, Van der Straeten D (2007) Monitoring and screening plant populations with combined thermal and chlorophyll fluorescence imaging. J Exp Bot 58:773–784

Chaerle L, Lenk S, Leinonen I, Jones HG, Van der Straeten D, Buschmann C (2009) Multi-sensor plant imaging: towards the development of a stress catalogue. Biotechnol J 4:1152–1167

Csefalvay L, Di Gaspero G, Matous K, Bellin D, Ruperti B, Olejnickova J (2009) Pre-symptomatic detection of *Plasmopara viticola* infection in grapevine leaves using chlorophyll fluorescence imaging. Eur J Plant Pathol 125:291–302

Delalieux S, van Aardt J, Keulemans W, Coppin P (2007) Detection of biotic stress (*Venturia inaequalis*) in apple trees using hyperspectral data: non-parametric statistical approaches and physiological implications. Eur J Agronomy 27:130–143

Dudareva N, Negre F, Nagegowda DA, Orlova I (2006) Plant volatiles: recent advances and future perspectives. Crit Rev Plant Sci 25:417–440

Franke J, Menz G (2007) Multi-temporal wheat disease detection by multi-spectral remote sensing. Precison Agric 8:161–172

Gebbers R, Adamchuk VI (2010) Precision agriculture and food security. Science 327:828–831

Hadjiloucas S, Walker GC, Bowen JW, Zafiropoulos A (2009) Propagation of errors from a null balance terahertz reflectometer to a sample's relative water content. J Phys: Conf Ser Sens Appl XV(178):012012, 1–5

Hillnhuetter C, Mahlein A-K, Sikora RA, Oerke E-C (2011a) Remote sensing to detect plant stress induced by *Heterodera schachtii* and *Rhizoctonia solani* in sugar beet fields. Field Crop Res 122:70–77

Hillnhuetter C, Mahlein A-K, Sikora RA, Oerke E-C (2011b) Use of imaging spectroscopy to discriminate symptoms caused by *Heterodera schachtii* and *Rhizoctonia solani* on sugar beet. Precision Agric 13:17–32

Hillnhuetter C, Sikora RA, Oerke E-C, van Dusschoten D (2012) Nuclear magnetic resonance: a tool for imaging belowground damage caused by *Heterodera schachtii* and *Rhizoctonia solani* on sugar beet. J Exp Bot 63:319–327

Huang LS, Zhao JL, Zhang DY, Dong YY, Zhang JC (2012) Identifying and mapping stripe rust in winter wheat using multi-temporal airborne hyperspectral images. Int J Agric Biol 14:697–704

Jacquemoud, S, Ustin SL (2001) Leaf optical properties: a state of the art. In Proceedings 8th international symposium physical measurements & signatures in remote sensing, CNES, Aussois, France, 8–12 Jan 2001, pp 223–232

Jones HG, Schofield P (2008) Thermal and other remote sensing of plant stress. Gen Appl Plant Physiol 34:19–32

Kobayashi T, Kanda E, Kitada K, Ishiguro K, Torigoe Y (2001) Detection of rice panicle blast with multispectral radiometer and the potential of using airborne multispectral scanners. Phytopathology 91:316–323

Kuckenberg J, Tartachnyk I, Noga G (2009) Temporal and spatial changes of chlorophyll fluorescence as a basis for early and precise detection of leaf rust and powdery mildew infections in wheat leaves. Precision Agric 10:34–44

Kushalappa AC, Lui LH, Chen CR, Lee B (2002) Volatile fingerprinting (SPMEGCFID) to detect and discriminate diseases of potato tubers. Plant Dis 86:131–137

Laothawornkitkul J, Moore JP, Taylor JE, Possell M, Gibson TD, Hewitt CN, Paul ND (2008) Discrimination of plant volatile signatures by an electronic nose: a potential technology for plant pest and disease monitoring. Environ Sci Tech 42:8433–8439

Lenthe J-H (2006) Erfassung befallsrelevanter Klimafaktoren in Weizenbeständen mit Hilfe digitaler Infrarot-Thermographie. PhD thesis, University of Bonn

Lenthe J-H, Oerke E-C, Dehne H-W (2007) Digital thermography for monitoring canopy health of wheat. Precision Agric 8:15–26

Li C, KrewerG, Kays SJ (2009) Blueberry postharvest disease detection using an electronic nose. ASABE paper no. 096783, ASABE annual international meeting, Reno, NV, June 21–June 24, 2009

Lindenthal M (2005) Visualisierung der Krankheitsentwicklung von Falschem Mehltau an Gurken durch *Pseudoperonospora cubensis* mittels Thermography. PhD thesis, University of Bonn

Lindenthal M, Steiner U, Dehne HW, Oerke EC (2005) Effect of downy mildew development on transpiration of cucumber leaves visualized by digital thermography. Phytopathology 95:233–240

Mahlein AK, Steiner U, Dehne HW, Oerke EC (2010) Spectral signatures of sugar beet leaves for the detection and differentiation of diseases. Precision Agric 11:413–431

Mahlein AK, Oerke EC, Steiner U, Dehne HW (2012a) Recent advances in sensing plant diseases for precision crop protection. Eur J Plant Pathol 133:197–203

Mahlein AK, Steiner U, Hillnhuetter C, Dehne HW, Oerke EC (2012b) Hyperspectral imaging for small-scale analysis of symptoms caused by different sugar beet diseases. Plant Methods 8:3. doi:10.1186/1746-4811-8-3

Mahlein AK, Rumpf T, Welke P, Dehne HW, Pluemer L, Steiner U, Oerke EC (2013) Development of spectral indices for detecting and identifying plant diseases. Remote Sens Environ 128:21–30

Markom MA, Shakaff AYM, Adom AH, Ahmad MN, Hidayat W, Abdullah AH, Fikri NA (2009) Intelligent electronic nose system for basal stem rot disease detection. Comput Electron Agric 66:140–146

Montes JM, Melchinger AE, Reif JC (2007) Novel throughput phenotyping platforms in plant genetic studies. Trends Plant Sci 12:433–436

Moshou D, Bravo C, West J, Wahlen S, McCartney A, Ramon H (2004) Automatic detection of 'yellow rust' in wheat using reflectance measurements and neural networks. Comput Electron Agric 44:173–188

Naidu RA, Perry EM, Pierce FJ, Mekuria T (2009) The potential of spectral reflectance technique for the detection of grapevine leafroll-associated virus-3 in two red-berried wine grape cultivars. Comput Electron Agric 66:38–45

Narvankar DS, Singh CB, Jayas DS, White NDG (2009) Assessment of soft X-ray imaging for detection of fungal infection in wheat. Biosyst Eng 103:49–56

Nutter F, van Rij N, Eggenberger SK, Holah N (2010) Spatial and temporal dynamics of plant pathogens. In: Oerke EC, Gerhards R, Menz G, Sikora RA (eds) Precision crop protection – the challenge and use of heterogeneity. Springer, Dordrecht, pp 27–50

Oerke EC (2006) Crop losses to pests. J Agric Sci 144:31–43

Oerke EC, Steiner U (2010) Potential of digital thermography for disease control. In: Oerke EC, Gerhards R, Menz G, Sikora RA (eds) Precision crop protection – the challenge and use of heterogeneity. Springer, Dordrecht, pp 167–182

Oerke EC, Steiner U, Dehne HW, Lindenthal M (2006) Thermal imaging of cucumber leaves affected by downy mildew and environmental conditions. J Exp Bot 57:2121–2132

Oerke EC, Gerhards R, Menz G, Sikora RA (2010) Preface. In: Oerke EC, Gerhards R, Menz G, Sikora RA (eds) Precision crop protection – the challenge and use of heterogeneity. Springer, Dordrecht

Oerke EC, Fröhling P, Steiner U (2011) Thermographic assessment of scab disease on apple leaves. Precision Agric 12:699–715

Pearson TC, Wicklow DT (2006) Detection of kernels infected by fungi. Trans ASABE 49 (4):1235–1245

Pietrzykowski E, Stone C, Pinkard E, Mohammed C (2006) Effects of *Mycosphaerella* leaf disease on the spectral reflectance properties of juvenile *Eucalyptus globules* foliage. Forest Pathol 36:334–348

Plaza A, Benediktsson JA, Boardman JW, Brazile J, Bruzzone L, Camps-Valls G et al (2009) Recent advances in techniques for hyperspectral image processing. Remote Sens Environ 113:110–122

Prithiviraj B, Vikram A, Kushalappa AC, Yaylayam V (2004) Volatile metabolite profiling for the discrimination of onion bulbs infected by *Erwinia carotovora* ssp. *carotovora*, *Fusarium oxysporum* and *Botrytis allii*. Eur J Plant Physiol 110:371–377

Quin J, Burks TF, Ritenour MA, Bonn WG (2009) Detection of citrus canker using hyperspectral reflectance imaging with spectral information divergence. J Food Eng 93:183–191

Reynolds GJ, Windels CE, MacRae IV, Laguette S (2012) Remote sensing for assessing *Rhizoctonia* crown and root rot severity in sugar beet. Plant Dis 96:497–505

Rousseau C, Belin E, Bove E, Fabre F, Berruyer R, Guillaumes J et al. (2013) High throughput quantitative phenotyping of plant resistance using chlorophyll fluorescence image analysis. Plant Methods 9:17 (http://www.plantmethods.com/content/9/1/17)

Rumpf T, Mahlein AK, Steiner U, Oerke EC, Dehne HW, Plümer L (2010) Early detection and classification of plant diseases with Support Vector Machines based on hyperspectral reflectance. Comput Electron Agric 74:91–99

Sankaran S, Mishraa A, Ehsania R, Davis C (2010) A review of advanced techniques for detecting plant diseases. Comput Electron Agric 72:1–13

Schmitz A, Kiewnick S, Schlang J, Sikora RA (2004) Use of high resolutional digital thermography to detect *Heterodera schachtii* infestation in sugar beets. Commun Agric Appl Biol Sci 69:359–363

Scholes JD, Rolfe SA (2009) Chlorophyll fluorescence imaging as tool for understanding the impact of fungal diseases on plant performance: a phenomics perspective. Funct Plant Biol 36:880–892

Spinelli F, Noferini M, Costa G (2006) Near infrared spectroscopy (NIRs): perspective of fire blight detection in asymptomatic plant material. In: Proceeding of 10th international workshop on fire blight. Acta Horticult 704:87–90

Stafford JV (2000) Implementing precision agriculture in the 21st century. J Agric Eng Res 76:267–275

Steddom K, Bredehoeft MW, Khan M, Rush CM (2005) Comparison of visual and multispectral radiometric disease evaluations of *Cercospora* leaf spot of sugar beet. Plant Dis 89:153–158

Steiner U, Buerling K, Oerke EC (2008) Sensorik für einen präzisierten Pflanzenschutz. Gesunde Pflanzen 60:131–141

Stenzel I, Steiner U, Dehne HW, Oerke EC (2007) Occurrence of fungal leaf pathogens in sugar beet fields monitored with digital infrared thermography. In: Stafford JV (ed) Precision Agriculture '07. Papers presented at the 6th European conference on precision agriculture. Wageningen Academic Publishers, pp 529–535

Stoll M, Schultz HR, Baecker G, Berkelmann- Loehnertz B (2008) Early pathogen detection under different water status and the assessment of spray application in vineyards through the use of thermal imagery. Precision Agric 9:407–417

Thenkabail PS, Smith RB, De Pauw E (2000) Hyperspectral vegetation indices and their relationship with agricultural crop characteristics. Remote Sens Environ 71:158–182

Vadivambal R, Jayas DS (2011) Applications of thermal imaging in agriculture and food industry – a review. Food Bioprocess Technol 4:186–199

Von Witzke H, Noleppa S, Schwarz G (2008) Global agricultural market trends and their impacts on European agriculture. Working paper 84, Humboldt University Berlin. http://www.agrar.hu-berlin.de/struktur/institute/wisola/publ/wp. Accessed 28 June 2011

Waggoner PE, Aylor DE (2000) Epidemiology, a science of patterns. Annu Rev Phytopathol 38:1–24

West JS, Bravo C, Oberti R, Lemaire D, Moshou D, McCartney HA (2003) The potential of optical canopy measurement for targeted control of field crop diseases. Annu Rev Phytopathol 41:593–614

West SJ, Bravo C, Oberti R, Moshou D, Ramon H, McCartney HA (2010) Detection of fungal diseases optically and pathogen inoculum by air sampling. In: Oerke EC, Gerhards R, Menz G, Sikora RA (eds) Precision crop protection – the challenge and use of heterogeneity. Springer, Dordrecht, pp 135–150

Zhang M, Qin Z, Liu X, Ustin S (2003) Detection of stress in tomatoes induced by late blight disease in California, USA, using hyperspectral remote sensing. Appl Earth Observation Geoinf 4:295–310

Part II
Case Studies and Special Applications

Diagnostic Challenges for the Detection of Emerging Pathogens: A Case Study Involving the Incursion of *Pseudomonas syringae* pv. *actinidiae* in New Zealand

Robert K. Taylor, Joanne R. Chapman, Megan K. Romberg, Bevan S. Weir, Joel L. Vanneste, Kerry R. Everett, Lisa I. Ward, Lia W. Liefting, Benedicte S.M. Lebas, and Brett J.R. Alexander

Abstract In November 2010, *Pseudomonas syringae* pv. *actinidiae* (Psa) was detected for the first time in New Zealand. This finding triggered one of the largest surveillance and diagnostic programmes seen in New Zealand's horticultural industry. During this response, over 912 kiwifruit orchards and 14,500 samples were screened and tested for the presence of Psa. The initial objectives of the response were to confirm the causal agent, determine disease prevalence and identify possible mechanisms of spread with the aim of identifying management options to contain the outbreak. Molecular diagnoses and characterisation of the Psa strains isolated during the response was conducted using a range of techniques that included qPCR,

R.K. Taylor (✉) • L.I. Ward • L.W. Liefting • B.S.M. Lebas • B.J.R. Alexander
Plant Health and Environment Laboratory, Ministry for Primary Industries, P.O. Box 2095, Auckland 1140, New Zealand
e-mail: robert.taylor@mpi.govt.nz

J.R. Chapman
Previously Plant Health and Environment Laboratory, Ministry for Primary Industries, P.O. Box 2095, Auckland 1140, New Zealand

Section for Zoonotic Ecology and Epidemiology, Linnaeus University, Kalmar SE-39182, Sweden

M.K. Romberg
Previously Plant Health and Environment Laboratory, Ministry for Primary Industries, P.O. Box 2095, Auckland 1140, New Zealand

USDA-APHIS, Beltsville, MD, USA

B.S. Weir
Landcare Research, Private Bag 92170, Auckland 1142, New Zealand

J.L. Vanneste
The New Zealand Institute for Plant and Food Research Ltd., Private Bag 3123, Hamilton 3240, New Zealand

K.R. Everett
The New Zealand Institute for Plant and Food Research Ltd., Private Bag 92169, Auckland 1025, New Zealand

M.L. Gullino and P.J.M. Bonants (eds.), *Detection and Diagnostics of Plant Pathogens*, Plant Pathology in the 21st Century 5, DOI 10.1007/978-94-017-9020-8_5,

rep-PCR fingerprinting, multilocus sequence analysis, and next generation sequencing. The usefulness and challenges of using the molecular techniques available at the time for Psa detection and characterisation during the response are discussed.

Keywords *Pseudomonas syringae* pv. *actinidiae* • *Actinidia* spp. • Diagnostics • Emerging pathogens • Response • Surveillance • New Zealand

Introduction

Pseudomonas syringae pv. *actinidiae* (Psa), the causal agent of bacterial canker of kiwifruit, is an emerging pathogen that has shown an increase in disease incidence and geographic range. Until recently, little had been published about the diagnostics, ecology and epidemiology of this pathogen. Prior to 2010, there were less than 15 peer-reviewed scientific papers on the diagnostics and biology of this pathogen (CAB abstracts, accessed 25-7-13). The following sections outline the available information on the geographic distribution, taxonomy, methods for identification and genetic diversity of Psa at the time of the detection in New Zealand in 2010, and the resulting diagnostic challenges that occurred during the response. In this chapter the term response refers to diagnostic, surveillance and management activities taken during the first 3 months after the Psa outbreak in New Zealand.

Geographic Distribution of Psa

Psa was first described on *Actinidia deliciosa* (green-fleshed kiwifruit) in Japan in 1984 (Takikawa et al. 1989) and subsequently was reported from Korea (Koh and Lee 1992), Italy (Scortichini 1994), and China (Liang et al. 2000) prior to the detection of Psa in New Zealand. In the last few years, Psa has spread rapidly throughout the majority of kiwifruit producing countries; it has been reported in Australia, Chile, France, New Zealand, Portugal, Spain, Switzerland and Turkey (Vanneste 2013). Subsequent to the first description of Psa in 1989, disease incidence and geographic spread of this pathogen was of relatively low incidence with the majority of countries reporting detections occurring after 2009. The increase in geographic spread coincided with the first detections of Psa on *A. chinensis* (gold-fleshed kiwifruit) in Italy in 2008 (Ferrante and Scortichini 2009), while previously it had only been found on *A. deliciosa*.

In Italy, Psa was first detected on *A. deliciosa* in 1992; however, epidemics of bacterial canker were not observed until 2008, at which time economic losses were estimated at around €2 million (Balestra et al. 2009). Worldwide, kiwifruit is important economically with production reaching up to 1.6 million tons/year and an estimated value of €56 billion worldwide (Balestra et al. 2009). In New Zealand, kiwifruit exports earn approximately $NZ 1.045 billion (€636 million) per year and are the second most important horticultural export crop by value (Fresh Facts 2012 http://www.freshfacts.co.nz/file/fresh-facts-2012.pdf).

The Taxonomy of Psa and Relationship to Other Bacterial Pathogens on Kiwifruit

Psa is a pathovar in the *Pseudomonas syringae* species complex. *Pseudomonas syringae* is the most economically important plant pathogenic species in the genus *Pseudomonas* and includes more than 60 pathovars. These pathovars cause numerous plant diseases with a variety of diverse symptoms including canker, dieback, blossom blight and leaf spot. Symptoms of Psa on kiwifruit include leaf spot, flower wilt, bacterial ooze, red exudate, canker and cane die-back, and can lead to vine death (Balestra et al. 2009). Psa is a gram-negative, motile, non-spore forming, rod-shaped bacterium. It is an obligate aerobe and elicits a hypersensitive reaction in tobacco (Takikawa et al. 1989).

The host range of Psa appears to be confined to species of *Actinidia* (Serizawa et al. 1989) and it has been recorded causing bacterial canker on *A. arguta, A. chinensis* and *A. deliciosa*. Psa has been shown to be weakly virulent to species of *Prunus* by artificial inoculation but has not been recorded to cause natural infection on stone fruit (Takikawa et al. 1989). DNA hybridisation studies on pathovars of *P. syringae* described nine unique genomic groups (Gardan et al. 1999) and Psa was placed in genomospecies 8 (Scortichini et al. 2002). Multilocus sequence typing confirmed previously described genomospecies and placed Psa in phylogroup 1 (Sarkar and Guttman 2004).

All bacterial pathogens recorded on kiwifruit prior to 2010 were species in the genus *Pseudomonas*. These include *P. syringae* pv. *syringae*, which was reported to cause canker and leaf spots Balestra and Varvaro (1997), and *Pseudomonas* sp. (formerly described as *P. viridiflava*) which causes blossom blight, bud rot and leaf spots (Young et al. 1997). In addition to these pathogenic species, *P. fluorescens* and *P. marginalis* also have been found to be associated with kiwifruit flowers as saprophytes (Everett and Henshall 1994). All of these species share cultural, morphological and/or pathogenic characters that are similar to those of Psa.

Methods for Identification of Psa

A definitive diagnosis of Psa can be obtained using a combination of biochemical, molecular and pathogenicity tests.

Takikawa et al. (1989) described colonies on nutrient agar plates as rounded, convex, glistening, translucent and creamy-white in colour and lacking a fluorescent pigment on King's medium B. Psa is gram-negative and exhibits characteristics of *P. syringae* LOPAT group 1a, i.e. levan positive, oxidase negative, potato soft rot negative, arginine dihydrolase negative and positive for tobacco hypersensitivity (Lelliott et al. 1966).

At the time of the outbreak in New Zealand, polymerase chain reaction (PCR) assays to identify Psa had been developed by Koh and Nou (2002) (primers

KNF/KNR), Rees-George et al. (2010) (primers Psa F1/R2 or Psa F3/R4), Sawada et al. (1997) (primers OCTF/OCTR), and Scortichini et al. (2002) (primers PAV1/P22). However, the KNF/KNR, OCTF/OCTR and PAV1/P22 primers (in some situations) would amplify DNA from other pseudomonads found on kiwifruit (Rees-George et al. 2010). However, sequencing of these amplified products can distinguish Psa from other leaf-spotting pseudomonads on kiwifruit (Park et al. 2011). The specificity of the primers PsaF1/R2 and PsaF3/R4 had been comprehensively tested (Rees-George et al. 2010; Vanneste et al. 2010) and were shown to amplify a 280 bp DNA fragment of 16S-23S rDNA origin from 37 authentic cultures of Psa. In contrast, this fragment was not amplified from 57 other strains of *Pseudomonas* (except for one strain of *P. syringae* pv. *theae*) or from 25 other species of plant pathogenic and saprophytic bacteria. Although the primers Psa F1/R2 and Psa F3/R4 did amplify *P. syringae* pv. *theae*, this pathogen has only been isolated from *Camellia sinensis* (tea plants) (Rees-George et al. 2010). These primers were recommended for the reliable identification of Psa but the authors observed that the level of detection may not be sufficient to allow the test to be used for routine screening of plants where inoculum levels maybe low (Rees-George et al. 2010). Identification of pure cultures can be confirmed by repetitive-PCR fingerprinting (Vanneste et al. 2010) and/or sequencing of the housekeeping genes citrate synthase (*cts*) (Vanneste et al. 2010), DNA gyrase B (*gyrB*), sigma factor 70 (*rpoD*) (Sarkar and Guttman 2004).

Genetic Diversity of Psa

Prior to the outbreak in New Zealand, genetic differences had been detected between Psa populations from different countries. Ferrante and Scortichini (2010) characterised a series of Psa strains isolated during epidemics of bacterial canker on *A. chinensis* in Italy by rep-PCR and multilocus sequence typing (MLST). This study showed that all strains isolated from the 2008–2009 Italian outbreaks all shared the same rep-PCR fingerprint and MLST profile but were different to the strains previously isolated in Japan (1984) and Italy (1994). In addition, none of the strains from the Italian 2008–2009 outbreaks possessed genes coding for the phytotoxins phaseolotoxin or coronatine, further differentiating these strains from those in Japan, Italy and Korea. The presence of these strains was confirmed by Vanneste et al. (2010); based on rep-PCR fingerprinting two distinct Psa strains from Japan, Korea and Italy were detected. Subsequent analysis of the *cts* gene of these strains consistently detected two haplotypes that differed by two base pairs; Psa strains from Japan, Korea and the Italian 1994 outbreak belonged to one haplotype whereas the Italian strains isolated from the epidemics in 2008–2009 belonged to another haplotype. These studies concluded that the epidemics in Italy during 2008–2009 appear to have been caused by a different Psa population than those previously recorded in Japan, South Korea and earlier Italian outbreaks.

Detection and Identification of Psa in New Zealand

Psa was first observed on *A. chinensis* in New Zealand in the Bay of Plenty region (Everett et al. 2011). This detection triggered one of the largest biosecurity responses seen in New Zealand's horticultural industry and lead to a joint biosecurity response initiated by the Ministry for Primary Industries (MPI) and the kiwifruit industry. The initial objectives of the response were to confirm the aetiology and disease prevalence, and identify possible mechanisms of spread with the aim of informing management options to contain the outbreak (Richardson et al. 2012).

Identification of the Causal Agent

In November 2010, *A. chinensis* vines exhibiting symptoms of extensive necrotic leaf spots, wilting and browning of flowers were observed on an orchard in Te Puke, Bay of Plenty, New Zealand. Samples were initially sent to Plant and Food Research (PFR) where they were tested for Psa by PCR using the primers Psa F1/R2 (Rees-George et al. 2010). Upon returning a positive PCR test result for Psa, PFR immediately notified MPI, New Zealand's national plant protection organisation. Samples were sent to MPI's Plant Health and Environment laboratory (PHEL) for confirmation and validation. *A. chinensis* leaves suspected of being infected with Psa were tested by PCR and isolations were attempted. Infected leaf tissue and bacterial colonies tested positive by PCR using the primers Psa F1/R2 and PAV1/P22. The amplified PCR products were sequenced and BLAST analysis showed that these nucleotide sequences were identical to reference Psa sequences in NCBI GenBank. Gram negative, white creamy bacterial colonies that were oxidase negative, positive for tobacco hypersensitivity, and did not produce a fluorescent pigment on King's media B were consistently isolated from both leaf spots and flowers. For further confirmation, the housekeeping genes *gyrB* and *rpoD* were amplified from these bacteria and phylogenetic analysis showed that the sequences of these genes were identical to those from the Psa type strain, placing the New Zealand strains into phylogroup 1 (Fig. 1). Verification of Psa as the plant pathogen required pathogenicity tests on kiwifruit seedlings (Everett et al. 2011).

Isolation of a culture and subsequent pathogenicity tests are still considered to be the gold standard for the definitive identification of bacterial pathogens. The timeframe for the detection and identification of Psa was relatively short ranging from a presumptive diagnosis at 8 h from receipt of sample, to a more definitive identification after 60 h, and completion of Koch's postulates 10 days later. However, detection of a pathogen of high-risk potential is time-sensitive when a response to an outbreak needs to be initiated. Biosecurity decision makers are faced with the challenge of balancing PCR-positive results with the time required to demonstrate viability and pathogenicity of the detected organism. This was

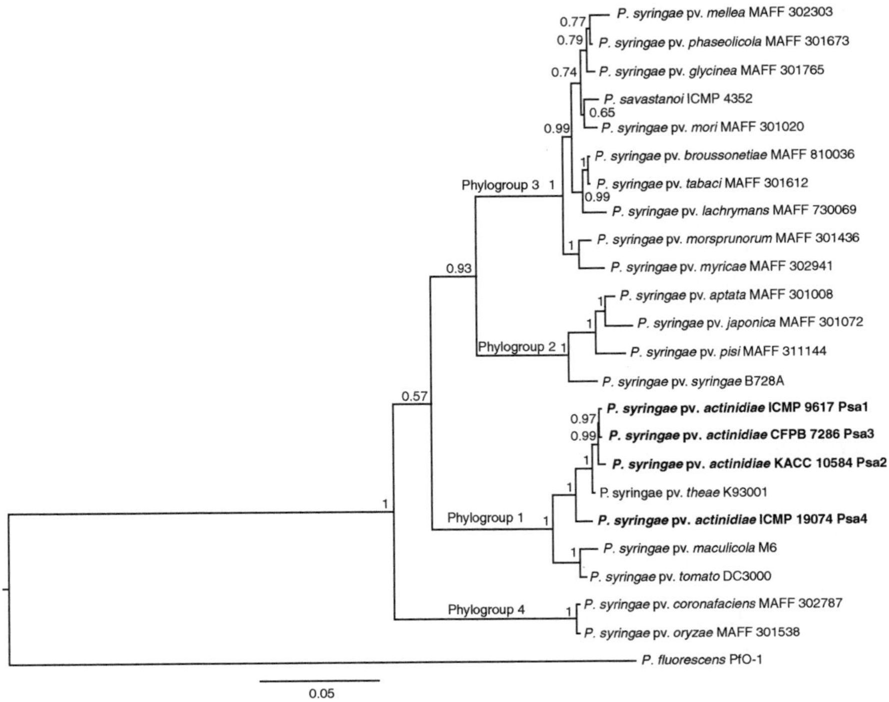

Fig. 1 Phylogenetic tree of selected *Pseudomonas syringae* pathovars showing their relationship to *P. syringae* pv. *actinidiae* derived from a concatenation of seven housekeeping genes (*acn1*, *cts*, *gapA*, *gyrB*, *pgi*, *rpoD*, *pfk* and pgi)

particularly the case for the first PCR detections of Psa on kiwifruit where there was a lack of additional validated diagnostic tools that could confirm the detection of Psa directly in diseased plant tissue.

Surveying New Zealand's Kiwifruit Orchards for Psa

Prior to the detection of Psa there had been no surveys or research conducted on bacterial diseases on kiwifruit in New Zealand for over 14 years. This resulted in a lack of baseline data on the background micro-flora present on kiwifruit and pre-existing disorders or leaf spotting diseases that could mask Psa symptoms and interfere with the diagnoses of the pathogen. Surveying for kiwifruit bacterial canker in the field proved to be challenging as the leaf spot symptoms observed at the outbreak site were found to be unreliable diagnostic indicators. Similar symptoms had been seen in New Zealand kiwifruit orchards for a number of years (Everett and Henshall 1994; Young et al. 1997) suggesting the possibility that Psa had been present in New Zealand for some time. Isolates of *Pseudomonas* species from kiwifruit that had been collected and stored in the International Collection of Microorganisms from Plants (ICMP) and at a PFR collection were screened by PCR to determine if Psa was already present in New Zealand. A total of 139 *Pseudomonas* isolates from ICMP and a further 52 *Pseudomonas* isolates from PFR tested negative for Psa by PCR (Park et al. 2011). The screening of these historical *Pseudomonas* isolates associated with kiwifruit suggested that Psa strains had not been isolated from New Zealand kiwifruit samples prior to the outbreak. Accessibility to these historic cultures demonstrates the utility and importance of maintaining a national collection of plant pathogenic micro-organisms. In addition, the vast collection of *Pseudomonas* isolates stored in the ICMP were of significant value for use as positive controls, optimisation and development of test methods and characterisation of Psa strains during the response.

Large-Scale Testing of Psa

Samples received during orchard surveys for Psa included kiwifruit budwood, canes, leaves, and pollen. The large number of samples that needed to be tested immediately required PHEL to scale up their procedures for the detection of Psa. The conventional PCR test to detect Psa had been previously validated for use in our laboratory. However, DNA extraction and PCR testing protocols for the detection of Psa in large quantities of samples had not been established. This required the immediate evaluation of DNA extraction methods and PCR techniques for high-throughput detection in kiwifruit plant material. From this evaluation the InviMag® DNA extraction kit (Stratec Molecular, Berlin, Germany) used in combination with a KingFisher automated extraction workstation (Thermo Scientific™, Waltham, MA, USA) was chosen for screening large samples because it consistently generated high quality DNA from leaf and pollen tissue and had a faster sample processing time than other methods tested (PHEL 2010 laboratory protocol). At the time of the initial detection a real-time PCR (qPCR) assay had not been developed for the detection of Psa. Therefore, PHEL rapidly developed a SYBR Green qPCR assay using the primers Psa F1/R2 to detect Psa in leaf and pollen tissue (PHEL 2010 laboratory protocol). The use of the DNA-binding dye SYBR green for the detection of PCR amplicons allowed for the rapid conversion of conventional to qPCR and enabled the high-throughput detection of Psa in a response setting. This approach proved to be more sensitive than using conventional PCR, reduced the risk of cross contamination or false positive results from PCR product carryover, allowed the specificity of the reaction to be further assessed on the basis of the melting temperature, and avoided the use of electrophoresis in the detection phase. Ultimately, the limiting steps for high-throughput testing was the sample preparation (debagging, labelling, and plant-tissue extraction) required for DNA extraction and the measures taken to monitor and minimise cross-contamination. Additionally, a series of controls to ensure reliable test results were included in each run which further reduces the amount of samples that can be processed per qPCR run. This includes a positive nucleic acid control, internal control, negative amplification control and a negative extraction control in each batch to monitor the potential for contamination during plant tissue and nucleic acid extraction. Once these controls were included and test samples were duplicated, only 22 samples could be tested in a 96-well qPCR plate. Despite this limitation, the above protocols processed approximately 200–300 samples per day. During the first 3 months of orchard surveillance, the laboratory tested samples from 912 properties and carried out multiple tests on approximately 14,500 leaf samples (Richardson et al. 2012).

Validation and Interpretation of qPCR Test Results

During the development of the qPCR test, limits of detection and cyclic threshold (Ct) cut-off values were evaluated by testing healthy kiwifruit tissue spiked with known concentrations of Psa. Based on these results it was determined that kiwifruit leaf tissue with reactions of a Ct value less than 30 were positive, between 30 and 35 were weak positive and Ct values greater than 35 were repeated. During the survey, the majority of kiwifruit leaf samples that tested positive by qPCR returned a result a Ct value between 16 and 30. Leaf samples that tested positive by qPCR were further confirmed by isolating a Psa culture from leaf spots. From these isolations, fluorescent and non-fluorescent colonies were obtained and tested positive for Psa by qPCR. This was initially concerning because the production of a fluorescent pigment in King's B was not recorded as a characteristic of Psa (Takikawa et al. 1989). It was therefore thought that the qPCR test maybe detecting other pseudomonads present on kiwifruit. However, further molecular testing and sequence analysis confirmed that these fluorescent isolates were also Psa (Everett et al. 2011). The detection of these atypical Psa cultures that produce fluorescence in King's B has now been reported for other isolates of Psa from Italy and France (Vanneste et al. 2011b, c).

Leaf samples that returned late amplification products with high Ct values greater than 30 were rare for infected leaf tissue at the start of the response. This likely was due to high numbers of Psa cells in newly infected leaves at that time of the year (late spring). Results of positive and negative test results from kiwifruit leaf tissue were further confirmed by isolation for Psa. Psa in most cases was readily isolated from leaf samples with positive qPCR results; whereas Psa was not isolated from tissue that returned negative PCR results. However, interpretation of late-amplification products that returned Ct values greater than 30 was challenging, especially when results could not be verified by isolating a culture. Melt curve analysis can be used to exclude false positive reactions that can arise due to non-specific amplification, and we used this to our advantage. There were a low number of positive test results that returned Ct values greater than 30 where Psa could not be isolated but the qPCR produced a specific melt curve and sequences of the qPCR product were identical to the Psa type strain. Interpretation of these results became problematic because it was discovered during the response that the primers Psa F1/R2 and PsaF3/R4 also amplify a PCR product identical to other pathovars of *P. syringae* (e.g. *P. syringae* pv. *passiflora*, *P. syringae* pv. *morsprunorum*). These pathovars have not been recorded to occur on kiwifruit leaf tissue; however, it did reduce public confidence in weak positive test results that could not be verified by other means. In particular, as *P. syringae* pv. *passiflora* occurs on passionfruit, some kiwifruit growers previously had grown passionfruit plants as a crop under their vines and were concerned with the potential for cross-contamination. Interpretation of high Ct values from SYBR green qPCR test results can be problematic but confidence in such results can be improved by the presence of a specific melt curve, sequencing of the qPCR product and further testing with another diagnostic assay.

qPCR Detection of Psa in Pollen

In New Zealand, kiwifruit pollen is harvested from closed flowers of *A. deliciosa* and is artificially applied to kiwifruit vines during flowering to enhance fruit size and consistency of pollination. Prior to the detection of Psa in New Zealand there were no published accounts of any association of phytopathogenic bacteria including Psa with kiwifruit pollen. However, there had been some work that year prior to the 2010 outbreak, during which Psa had been isolated from pollen collected from open flowers in an Italian orchard (later published in Vanneste et al. 2011a).

Testing of pollen showed that known amounts of Psa bacteria co-extracted with pollen tissue caused amplification curves to occur later than those observed for leaf samples. This was likely due to the presence of PCR inhibitory substances found in pollen. Therefore, qPCR Ct values were adjusted for pollen samples, a Ct value less than 37 were considered positive; between 37 and 42 were weak positives and greater than 42 were negative. During the survey pollen samples (milled) were collected from throughout New Zealand and tested by qPCR for the presence of Psa. Low levels of Psa were detected in 95 % of the pollen samples tested. Positive test results produced Ct values between 30 and 39, melt curves were identical to those of Psa and sequencing of the PCR products were identical to those of the type strain of Psa. Attempts were then made to retrieve viable Psa colonies from pollen suspensions by plating out onto bacteriological media. The resulting colony growth on the plates were then re-suspended in saline and tested directly by qPCR. These bacterial suspensions tested positive for Psa indicating the detection of viable colonies. However, the isolation of a Psa culture was unsuccessful despite screening of hundreds of colonies that grew out of pollen suspensions. This may have been due to the PCR detecting dead cells or to low concentrations of Psa being outgrown by other bacteria. Consequently, it was not possible to conclusively determine if Psa detected in pollen was viable or to validate the test results by other methodology. These detections raised concerns that the qPCR was detecting dead cells or false positives by cross-reacting with other bacteria present in pollen. Interpretation and confidence in these test results became even more challenging when pollen batches from 2007 and pollen samples collected from areas with no Psa symptoms tested positive for Psa.

Based on these results, the movement of pollen and artificial pollination was restricted that year. Subsequent work in the following year using semi-selective media and qPCR confirmed that viable Psa colonies were present in New Zealand pollen collected from symptomatic and asymptomatic kiwifruit vines (MPI report 2011). Vanneste et al. (2011a) had obtained similar results by isolating Psa from pollen sourced from Italian orchards where no symptoms of bacterial canker had been detected. These results had important implications since the collection of pollen from kiwifruit blocks that appear visually free of disease may not be sufficient to ensure that harvested pollen is free of Psa. Whether Psa is readily disseminated by artificial pollination still remains to be determined. Initial epidemiology work by Richardson et al. (2012) did not find a link between pollen application and subsequent Psa infection.

Molecular Characterisation of Psa: Identifying the Outbreak Strain

The identification of Psa as the causal agent of the kiwifruit disease was quickly obtained very early during the response investigation; however, the identification of the outbreak strain was more challenging. As described in section "Genetic diversity of Psa", genetic differences had been detected among overseas Psa strains and the questions was raised whether the New Zealand strain was similar to those strains causing an epidemic of bacterial canker in Italy.

Molecular-Typing of Psa Strains Present in New Zealand

Initially, surveys suggested that Psa was widespread throughout kiwifruit growing regions in New Zealand with detections on both the North and South Islands. However, symptoms from the Bay of Plenty in the North Island were distinct; leaf spots, cane die-back, wilt, and canker were observed whereas only leaf spotting was observed in all other regions on both islands. Only after molecular analysis of strains isolated from different geographic regions in New Zealand was it revealed that two distinct strains were present; one causing more aggressive symptoms (Chapman et al. 2012; Vanneste et al. 2013). This was based on a combination of *cts* haplotype, rep-PCR fingerprinting (Vanneste et al. 2013), and multilocus sequence analysis of housekeeping, type III effector and phytotoxin genes (Chapman et al. 2012). The New Zealand Psa isolate was shown to be very similar to Psa strains causing bacterial kiwifruit canker in Italy, China, and Chile

(Chapman et al. 2012; Vanneste et al. 2013). This strain was recently classified as Psa biovar 3 (previously known as Psa V) in New Zealand (Vanneste et al. 2013). Field observations showed that the Psa biovar 3 was highly virulent and associated with leaf spot and canker symptoms; whereas, Psa biovar 4 (previously known as Psa LV) was only found to be associated with leaf spots (Vanneste et al. 2013). This link with virulence was further confirmed by pathogenicity tests (Vanneste et al. 2013). These results then limited the focus of response to biovar 3. Results from further surveys determined that biovar 3 was not as widespread throughout New Zealand as initially thought and mainly present in the Bay of Plenty, North Island.

At the time of the surveys, strain determination required a pure culture a process that could take up to 7 days followed by PCR, and rep-PCR fingerprinting and/or DNA sequencing. This process delayed results significantly and hampered efforts to survey for the pathogen. Rapid qPCR protocols recently have been developed to distinguish or determine Psa isolates without the need to culture the bacterium for DNA sequencing (E. Rikkerink; J. F. Mackay, personal communication).

Next-Generation Sequencing

The advent of high throughput sequencing has made it feasible to determine the genome sequence for phytopathogenic bacteria in a short time and at a cost equivalent to conducting a MLSA study. Thus, the original outbreak strain was sequenced using Roche 454 GS Junior sequencing platform and within the first weeks of the outbreak a draft genome sequence was assembled, in 3 days for a few thousand New Zealand dollars. Although draft genome sequences maybe prone to errors and gaps, it did allow analysis of effector and toxin genes known to be key virulence determinants in the *P. syringae* complex. A local BLAST database of the contigs was queried with these reference gene sequences and the presence or absence of these genes differed between Psa outbreak strains (Table 1). This analysis of the draft genome sequence quickly validated the identification of the New Zealand Psa isolate and provided significant insights into possible strain type. The draft sequence enabled key effector and phytotoxin genes to be screened and revealed some commonalties with the New Zealand Psa strain and the more aggressive Italian strain. The key differences found between the historic Psa strains (Japan 1989, Italy 1994) and the recent New Zealand and Italy outbreak strains was that they shared the effector *hopA1* gene and lacked genes for coronatine and phaseolotoxin production.

Although the draft New Zealand Psa genome sequence provided important information for Psa diagnoses and strain identification, conclusive data could not be obtained due to the lack of other Psa genomes for comparison. At the time, the only other Psa genome sequence available was that of the Japanese Psa type strain Kw 11, which was partially completed and there were no other genome sequences available for the epidemic causing Psa strains. If these sequences had been

Table 1 Comparison of the presence (+) or absence (−) of effector and toxin genes of Psa outbreak strains

Gene	Japan 1989	Italy 1994	Italy 2008/2009	New Zealand 2010
avrPto1	−	−	−	−
avrD1	+	+	+	+
avrAE1	+	+	+	+
hopA1	−	−	+	+
hopB1	−	−	−	−
hopC1	−	−	−	−
hopD1	+	+	+	+
hopF2	−	−	−	−
hopG1	−	−	−	−
hrpK1	+	+	+	+
hopAF1	+	−	−	−
hopAN1	+	+	+	+
Coronatine	+	−	−	−
Phaseolotoxin	+	+	−	−

available, significant advances could have been made in characterisation of New Zealand Psa strains and diagnostics for the detection and tracking of Psa during the outbreak. It was not until several months later that PCR diagnostics were developed to differentiate Psa biovars based on genome sequence data from a number of isolates from different geographic regions sequenced by PFR (E. Rikkerink, personal communication).

Conclusions

The introduction of Psa into New Zealand underscored a number of challenges in the use of molecular diagnostics to detect and characterise exotic phytopathogenic bacteria. The ideal situation during a response to a new outbreak is that the epidemiology of the pathogen are well understood and that reliable and specific diagnostic assays are available. At the time of the outbreak there were significant knowledge gaps in the biology of Psa and a lack of validated diagnostic tools for the high throughput detection of Psa in kiwifruit plant material. The timeframe to validate the identification of Psa was further hampered by the need to culture the causal agent and to conduct pathogenicity tests to demonstrate Koch's postulates.

Surveys conducted during the response illustrated the importance of continued crop surveillance. There was a lack of recent data on the microbial diversity present on kiwifruit, including other bacteria that may cause leaf spotting or that may interfere with the detection of Psa. Screening pseudomonad isolates in the ICMP collection and a laboratory collection from a study of pseudomonads on kiwifruit (Everett and Henshall 1994) confirmed that Psa was not present in New Zealand before the outbreak (Park et al. 2011). Additionally, a high microflora of pseudomonads present on kiwifruit leaves complicated the accurate detection and identification of Psa strains, in particular Psa biovar 3. Many of these pseudomonads share similar morphological and molecular characteristics to Psa. Most importantly, this included Psa biovar 4, formerly known as Psa LV, a weak pathogen and has been speculated to have remained undetected in New Zealand for many years previous to the outbreak (Vanneste et al. 2013).

Another challenge that many diagnosticians face is the shift from small sample loads to high-throughput testing. The requirement to test hundreds of samples used significant resources to rapidly adapt and validate the conventional PCR assay to a SYBR green qPCR protocol. These steps removed resources from other response activities and ultimately delayed test results. This demonstrates the need for continual improvement of diagnostic protocols for the detection of exotic and emerging organisms.

The interpretation of molecular test results without confirmation by a pure culture is a challenge for any diagnostician and proved as such in this outbreak. In our instance interpretation of qPCR test results with borderline Ct values proved challenging when the results could not be verified by a Psa culture. This was more problematic for detection of Psa in pollen because a selective medium was not available. The isolation of Psa from pollen can be challenging because presumptive Psa colonies can be morphologically similar to numerous saprophytes that are present in pollen.

Recent advances in Psa diagnostics using data from whole genome sequences has resulted in additional specific qPCR tests to distinguish between the Psa biovars (Balestra et al. 2013; Gallelli et al. 2014; Rikkerink et al. 2011; J. F. Mackay, dnature Limited). One of these tests target Psa biovar 3 and includes duplexing an internal control in the same reaction which has increased high-throughput detection, reduced test costs and is commercially available (dnature Limited, Gisborne, New Zealand). A next generation sequencing approach quickly produced a draft Psa genome sequence from the outbreak strain and provided important information to validate identification and assist with strain characterisation. At the time there was only one other draft genome sequence available for comparative analysis, now there are over 40 Psa genome sequences from multiple geographic origins. Had this data been available at the time of the outbreak in New Zealand it would have provided immediate results for strain characterisation providing insights into origin and pathogenic type (Butler et al. 2013; Mazzaglia et al. 2012; McCann et al. 2013). In the future, this could lead to the development of molecular tests that identify unique markers that detect pathogenic strain types and offer a possible alternative to pathogenicity testing.

If the recent developments in Psa diagnostics and strain characterisation had been available at the start of the outbreak, it would have avoided the need for assay development, yielded higher quality data, provided increased confidence levels in test results, and allowed more resources to survey and track the pathogen. As of September 2013, 76 % (2,265 orchards) of New Zealand's kiwifruit hectares are on orchards where Psa biovar 3 has been identified (Kiwifruit Vine Health http://www.kvh.org.nz/statistics). A recent study has indicated that Psa is expected to cost the New Zealand kiwifruit industry between $310 and $410 million over the next 5 years (Greer and Saunders 2012).

Acknowledgement The diagnostic development and activities undertaken during this response involved a large number of people. The authors would like to recognise the efforts of all the scientists and technical staff at PHEL, Plant and Food Research, Landcare, colleagues in MPI, Zespri, KVH, AsureQuality and kiwifruit industry staff who were involved in the initial response.

References

Balestra GM, Varvaro L (1997) *Pseudomonas syringae* pv. *syringae* causal agent of disease on floral buds of *Actinidia deliciosa* (A. Chev) Liang et Ferguson in Italy. J Phytopathol 145:375–378

Balestra GM, Mazzaglia A, Quattrucci A, Renzi M, Rossetti A (2009) Current status of bacterial canker spread on kiwifruit in Italy. Australas Plant Dis Notes 4:34–36

Balestra GM, Taratufolo MC, Vinatzer BA, Mazzaglia A (2013) A multiplex PCR assay for detection of *Pseudomonas syringae* pv. *actinidiae* and differentiation of populations with different geographic origin. Plant Dis 97(4):472–478

Butler MI, Stockwell PA, Black MA, Day RC, Lamont IL, Poulter RT (2013) *Pseudomonas syringae* pv. *actinidiae* from recent outbreaks of kiwifruit bacterial canker belong to different clones that originated in China. PLoS One 8(2):e57464

Chapman J, Taylor RK, Weir BS, Romberg MK, Vanneste JL, Alexander BJR (2012) Phylogenetic relationships of global *Pseudomonas syringae* pv. *actinidiae* populations isolated from green (*Actinidia deliciosa*) and gold (*A. chinensis*) kiwifruit vines. Phytopathology 102:1034–1044

Everett KR, Henshall WR (1994) Epidemiology and population ecology of kiwifruit blossom blight. Plant Pathol 43:824–830

Everett KR, Rees-George JRG, Taylor RK, Romberg MR, Vanneste JL, Manning MA, Fullerton R (2011) First report of *Pseudomonas syringae* pv. *actinidiae* causing kiwifruit bacterial canker in New Zealand. Australas Plant Dis Notes 6: 67–71

Ferrante P, Scortichini M (2009) Identification of *Pseudomonas syringae* pv. *actinidiae* as causal agent of bacterial canker of yellow kiwifruit (*Actinidia chinensis* Planchon) in central Italy. J Phytopathol 157:768–770

Ferrante P, Scortichini M (2010) Molecular and phenotypic features of *Pseudomonas syringae* pv. *actinidiae* isolated during recent epidemics of bacterial canker on yellow kiwifruit (*Actinidia chinensis*) in central Italy. Plant Pathol 59:954–962

Gallelli A, Talocci S, Pilotti M, Loreti S (2014) Real-time and qualitative PCR for detecting *Pseudomonas syringae* pv. *actinidiae* isolates causing recent outbreaks of kiwifruit bacterial canker. Plant Pathol 63(2):264–276

Gardan L, Shafik H, Belouin S, Broch R, Grimont F, Grimont PAD (1999) DNA relatedness among the pathovars of *Pseudomonas syringae* and description of *Pseudomonas* tremae sp. nov. and *Pseudomonas cannabina* sp. nov.(ex Sutic and Dowson 1959). Int J Syst Bacteriol 49:469–478

Greer G, Saunders C (2012) The costs of Psa-V to the New Zealand kiwifruit industry and the wider community. Agribusiness and Economics Research Unit Report, Lincoln University, p 62

Koh Y, Lee D (1992) Canker of kiwifruit by *Pseudomonas syringae* pv. *morsprunorum*. Korean J Plant Pathol 8:119–122

Koh YJ, Nou IS (2002) DNA markers for identification of *Pseudomonas syringae* pv. *actinidiae*. Mol Cells 13:309–314

Lelliott RA, Billing E, Hayward AC (1966) A determinative scheme for the fluorescent plant pathogenic pseudomonads. J Appl Bacteriol 29(3):470–489

Liang Y, Zhang X, Tian C, Gao A, Wang P (2000) Pathogenic identification of kiwifruit bacterial canker in Shaanxi. J Northwest For Coll 15(1):37–39

Mazzaglia A, Studholme DJ, Taratufolo MC, Cai R, Almeida NF, Goodman T, Balestra GM (2012) *Pseudomonas syringae* pv. *actinidiae* isolates from recent bacterial canker of kiwifruit outbreaks belong to the same genetic lineage. PLoS One 7(5):e36518

McCann HC, Rikkerink EH, Bertels F, Fiers M, Lu A, Rees-George J, Andersen MT, Gleave AP, Haubold B, Wohlers MW, Guttman DS, Wang PW, Straub C, Vanneste J, Rainey PB, Templeton MD (2013) Genomic analysis of the kiwifruit pathogen *Pseudomonas syringae* pv. *actinidiae* provides insight into the origins of an emergent plant disease. PLoS Pathog 9(7):e1003503

Ministry for Primary Industries Report (2011) Detection of *Pseudomonas syringae* pv. *actinidiae* from leaves and pollen collected from symptomatic and asymptomatic *Actinidia chinensis* in Te Puke, Bay of Plenty. http://www.kvh.org.nz/vdb/document/91512

Park D, Than D, Everett KR, Rees-George J, Romberg MK, Alexander BJR (2011) Screening of historical isolates stored in New Zealand culture collections for *Pseudomonas syringae* pv. *actinidiae*. New Zeal Plant Prot 64:292

Rees-George J, Vanneste JL, Cornish DA, Pushparajah IPS, Yu J, Templeton MD, Everett KR (2010) Detection of *Pseudomonas syringae* pv. *actinidiae* using polymerase chain reaction (PCR) primers based on the 16S–23S rDNA intertranscribed spacer region and comparison with PCR primers based on other gene regions. Plant Pathol 59:453–464

Richardson E, McFadden A, Rawdon T (2012) Initial outbreak investigations of *Pseudomonas syringae* pv. *actinidiae* in kiwifruit in New Zealand. Surveillance 39:36–42

Rikkerink E, Andersen MT, Rees-George J, Cui W, Vanneste J, Templeton MD (2011) Development of a rapid tool for the molecular characterisation of Psa haplotypes. Plant and food research report to Zespri Group Limited Ref. VI1256. SPTS no. 6361

Sarkar SF, Guttman DS (2004) Evolution of the core genome of *Pseudomonas syringae*, a highly clonal, endemic plant pathogen. Appl Environ Microbiol 70:1999–2012

Sawada H, Takeuchi T, Matsuda I (1997) Comparative analysis of *Pseudomonas syringae* pv. *actinidiae* and pv. *phaseolicola* based on phaseolotoxin-resistant ornithine carbamoyltransferase gene (argK) and 16S-23S rRNA intergenic spacer sequences. Appl Environ Microbiol 63:282–288

Scortichini M (1994) Occurrence of *Pseudomonas syringae* pv. *actinidiae* on kiwifruit in Italy. Plant Pathol 43:1035–1038

Scortichini M, Marchesi U, Di Prospero P (2002) Genetic relatedness among *Pseudomonas avellanae*, *P. syringae* pv. *theae* and *Ps* pv. *actinidiae*, and their identification. Eur J Plant Pathol 108:269–278

Takikawa Y, Serizawa S, Ichikawa T, Tsuyumu S, Goto M (1989) *Pseudomonas syringae* pv. *actinidiae* pv. nov.: the causal bacterium of canker of kiwifruit in Japan. Ann Phytopathol Soc Jpn 55:437–444

Vanneste JL (2013) Recent progress on detecting, understanding and controlling *Pseudomonas syringae* pv. *actinidiae*: a short review. New Zeal Plant Prot 66:170–177

Vanneste JL, Yu J, Cornish DA (2010) Molecular characterisations of *Pseudomonas syringae* pv. *actinidiae* strains isolated from the recent outbreak of bacterial canker on kiwifruit in Italy. New Zeal Plant Prot 63:7–14

Vanneste JL, Giovanardi D, Yu J, Cornish DA, Kay C, Spinelli F, Stefani E (2011a) Detection of *Pseudomonas syringae* pv. *actinidiae* in kiwifruit pollen samples. New Zeal Plant Prot 64:241–245

Vanneste JL, Kay C, Onorato R, Yu J, Cornish DA, Spinelli F, Max S (2011b) Recent advances in the characterisation and control of *Pseudomonas syringae* pv. *actinidiae*, the causal agent of bacterial canker on kiwifruit. Acta Hortic 913:443–455

Vanneste JL, Yu J, Cornish DA, Max S, Clark G (2011c) Presence of *Pseudomonas syringae* pv. *actinidiae*, the causal agent of bacterial canker of kiwifruit, on symptomatic and asymptomatic tissues of kiwifruit. New Zeal Plant Prot 64:241–245

Vanneste JL, Yu J, Cornish DA, Tanner DJ, Windner R, Chapman JR, Taylor RK, Mackay JF, Dowlut S (2013) Identification, virulence, and distribution of two biovars of *Pseudomonas syringae* pv. *actinidiae* in New Zealand. Plant Dis 97:708–719

Young JM, Gardan L, Ren XZ, Hu FP (1997) Genomic and phenotypic characterization of the bacterium causing blight of kiwifruit in New Zealand. Plant Pathol 46:857–864

Detection of Human Pathogens on Plants

Li Maria Ma, Jacqueline Fletcher, and Guodong Zhang

Abstract Unlike most plant pathogens, which typically multiply to large numbers after colonizing tissues of susceptible plants, human pathogens that associate with plants often fail to thrive in this environment and usually occur in low numbers. Nevertheless, their presence on plants could have significant public health and economic consequences. In recent years, national and international disease outbreaks associated with human pathogens on plant products, such as lettuce, spinach, green onions, seeds, sprouts, peppers, spices, tomatoes, and cantaloupes, have occurred frequently. Current standardized assays for the detection of major human pathogens on plants rely largely on microbiological, biochemical, and immunological analyses that are laborious and time consuming. Newer molecular-based methods, such as PCR, loop mediated isothermal amplification, and metagenomics approaches offer enhanced speed and sensitivity, and some of these have already been incorporated into the standard assays. However, molecular detection methods do not produce a live microbial isolate, which may be needed for government regulatory actions and future scientific studies. New enrichment strategies (especially the use of chromogenic selective media) have made culture detection more sensitive and accurate. Effective detection and diagnostic methods of the future will continue to differ in features depending upon the intended application and operators.

L.M. Ma (✉)
Department of Entomology and Plant Pathology, National Institute for Microbial Forensics & Food and Agricultural Biosecurity, Oklahoma State University, Stillwater, OK 74078, USA
e-mail: li.ma@okstate.edu

J. Fletcher
Department of Entomology & Plant Pathology and National Institute for Microbial Forensics & Food and Agricultural Biosecurity, Stillwater, OK 74078, USA

G. Zhang
Center for Food Safety and Applied Nutrition, Food and Drug Administration, College Park, MD 20174, USA

M.L. Gullino and P.J.M. Bonants (eds.), *Detection and Diagnostics of Plant Pathogens*, Plant Pathology in the 21st Century 5, DOI 10.1007/978-94-017-9020-8_6,

Keywords Detection • Human pathogens • Plant • Salmonella • Shiga toxin-producing *E. coli* • Listeria monocytogenes

Human Pathogens on Plants

Plant pathologists are concerned primarily about microbes that cause plant diseases, negatively impacting production, quality and trade. However, certain human pathogens, especially enteric microbes such as pathogenic *Escherichia coli* and *Salmonella* spp., also contaminate, colonize and even invade plants. Human diseases caused by such plant contaminants are becoming more common, widespread, and consequential, and national food safety agencies across the globe are seeking greater understanding of the mechanisms and interactions of human pathogens on plants (HPOPs). An increasing number of plant pathologists, realizing that there are significant similarities (and differences) among the two microbial groups and their interactions on plants, have intensified their research efforts in this area (Fletcher et al. 2013). Furthermore, thanks to increasing opportunities for interactions, collaborations and cooperation among plant pathologists and food microbiologists, a cross-disciplinary synergy has developed from which novel, robust and sustainable solutions to HPOP challenges will emerge.

Fresh produce has been associated repeatedly, and with increasing frequency, with outbreaks of foodborne illnesses (Lynch et al. 2009; Sivapalasingam et al. 2004). Leafy greens, melons, sprouts, berries, tomatoes and green onions, often eaten with little or no processing steps to eliminate pathogens, are among the most common produce implicated. Shiga toxin-producing *Escherichia coli* O157:H7 has been found on leafy greens (Hilborn et al. 1999; Wendel et al. 2009), *Salmonella* spp. on tomatoes, peppers and cantaloupes (Behravesh et al. 2011; Bowen et al. 2006; Gupta et al. 2007; Mody et al. 2011), hepatitis A virus on green onions (Wheeler et al. 2005), Shiga toxin-producing *E. coli* O104 on fenugreek seed sprouts (Danhorn and Fuqua 2007) and *Listeria monocytogenes* on cantaloupe (CDC 2011). In many cases contamination occurs either in the field or in the processing phase. Many of the common HPOPs colonize the intestinal tracts of vertebrates such as cattle and birds (including poultry) without causing symptoms in these hosts, and may affect humans only incidentally. Many HPOPs also exist in environments where plants are grown. When illness outbreaks affect significant numbers of people, leading to recalls of implicated produce, the economic impact can reach not only growers but also produce companies, processors, packers, distributors, retail stores, and the general public. In a single year (September 2011–September 2012) the FDA issued recalls of 56 produce items including fresh-cut fruit and vegetables and bagged vegetables that contained *Listeria* spp., pathogenic *E. coli* or *Salmonella* (U.S. FDA 2013b).

Opportunities for HPOP contamination of fresh produce begin on the farm and continue through all nodes of the food production and distribution chain, not ending until the food is consumed (Fig. 1). How pathogens move, directly or indirectly,

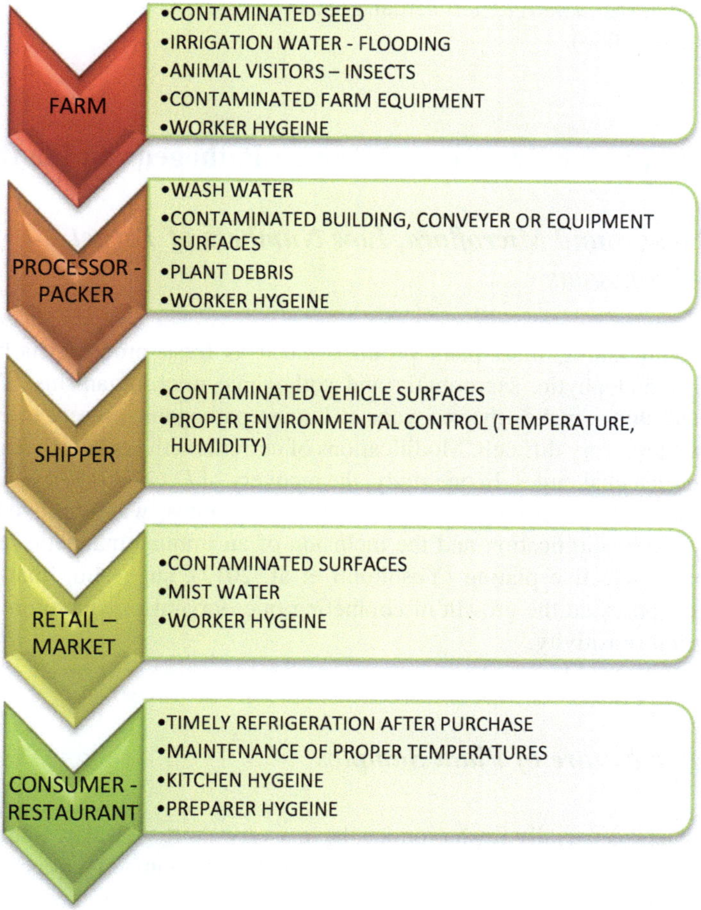

Fig. 1 Some potential points of microbial contamination of plant-derived foods, farm to fork

from vertebrate sources into plant foods can be complex and multi-faceted (Brandl 2006; Barak and Schroeder 2012; Fletcher et al. 2013; Teplitski et al. 2009; Twardoń et al. 2005). Understanding these sources and pathways is critical for the development of prevention and mitigation strategies.

Interestingly, the Gram negative bacterial family Enterobacteriaceae, which includes many of the human pathogens associated with plant foods (e.g., *Escherichia, Salmonella, Shigella*), also contains several plant pathogens (*Enterobacter, Erwinia, Pantoea, Pectobacterium,* etc.). The taxonomic relatedness of these plant and human pathogens raises interesting questions about the possibilities for niche competition or synergism, horizontal gene exchange in protected plant niches, or even host range expansion. A few cross-kingdom pathogens such as *Pseudomonas aeruginosa*, *Burkholderia cepacia*, *Dickeya* spp., *Enterococcus faecalis* and *Serratia marcescens* actually cause disease on both plants and humans (Vidaver et al. 2006).

Special Issues for Detection of Human Pathogens on Plants

High Background Microflora, Low Numbers of Target Human Pathogens

As raw commodities, most plant products, such as fresh produce, harbor large numbers of endophytic, saprophytic and pathogenic microorganisms. This high background clutter makes the recovery of human pathogens, present in comparatively low titers, very difficult. Modifications of current methods have been made to overcome such challenges. In one study, the recovery of *E. coli* O157:H7 from 24-h enrichment broths prepared from the plant host, cilantro, was improved significantly by broth acidification and the inclusion of an immunomagnetic separation step prior to selective plating (Yoshitomi et al. 2012), suggesting that the acid treatment suppressed the growth of competing microorganisms, therefore increasing detection sensitivity.

Perishable Nature of Plant Samples

Plant products, especially fresh produce, have a very short shelf life. Rapid on-site or in-field assays and platforms are needed for detecting human pathogens in these perishable samples, and efforts have been made to reduce assay time. Combining enrichment with real-time PCR provided for rapid (1 day) and sensitive detection of human pathogens on plants (Delibato et al. 2013). Detection of *Salmonella* on smooth tomato surfaces (10 CFU/tomato) within 8 h of inoculation was achieved by combining brief enrichment, flow-through immunocapture (FTI), and real-time PCR (Warren et al. 2007). Biosensors (described below) could provide additional solutions.

Nonhomogeneous Distribution of Target Pathogens in Plant Samples

Many plant products are difficult to homogenize, so extraction of target pathogen DNA can be challenging (Chen et al. 2011). Among a group of food matrices including whole milk, soft cheese, turkey deli meat, smoked salmon and alfalfa sprouts, the latter were the most difficult to process, and assay sensitivity for inoculated human pathogens was negatively impacted. Using irrigation water (applied to raw alfalfa sprouts) or rinse water (from leafy greens), instead of the raw commodities themselves, for assessing the presence of human pathogens may give a more accurate representation of the contamination level (Johnston et al. 2005).

Moving Target

Some foodborne human pathogens can be a 'moving target' for detection. For example, a Shiga toxin-producing *E. coli* O104:H4 associated with fenugreek sprouts caused a large and severe outbreak of foodborne illness in Germany in 2011 (Mellmann et al. 2011; Bielaszewska et al. 2011). This strain was an atypical Shiga toxin-producing *E coli* (STEC) resulting from horizontal gene transfer events between two distinct pathogenic *E. coli* groups: enteroaggregative and enterohemorrhagic (EAEC and EHEC) (Mellmann et al. 2011). This microbe would have been missed in the current standard STEC detection assay, which focuses on typical STEC serotypes: O157, O26, O45, O103, O111, O121, and O145 (U.S. FDA 2013a; USDA 2013). Many of the virulence associated genes in foodborne human pathogens are located on exchangeable genetic elements, such as plasmids, bacteriophages, and transposons, which can take part in horizontal gene transfer, resulting in the creation of new and dangerous bacterial strains. Therefore, it has been proposed recently to establish standardized genetic tests targeting virulence features that are easily exchangeable among bacterial strains (such as Shiga toxin genes) to improve the detection of rare and unusual strains of foodborne human pathogens (Bloch et al. 2012).

Detection and Predicting Infectivity

The use of more sensitive and rapid methods for the detection of HPOPs generates a dilemma in assessing the risk to human health associated with samples that test positive. Harsh conditions in the pre-harvest field production environment and post-harvest disinfection treatments may have an effect on microbial viability. Positive results from molecular detection assays may overestimate the actual number of infectious cells as well as the likelihood that human illness could ensue.

Detection of Human Pathogens on Plants

Standard Assays (Culture-Based Methods)

In the United States, the food safety system includes two major regulatory agencies: the U.S. Food and Drug Administration (FDA) and the U.S. Department of

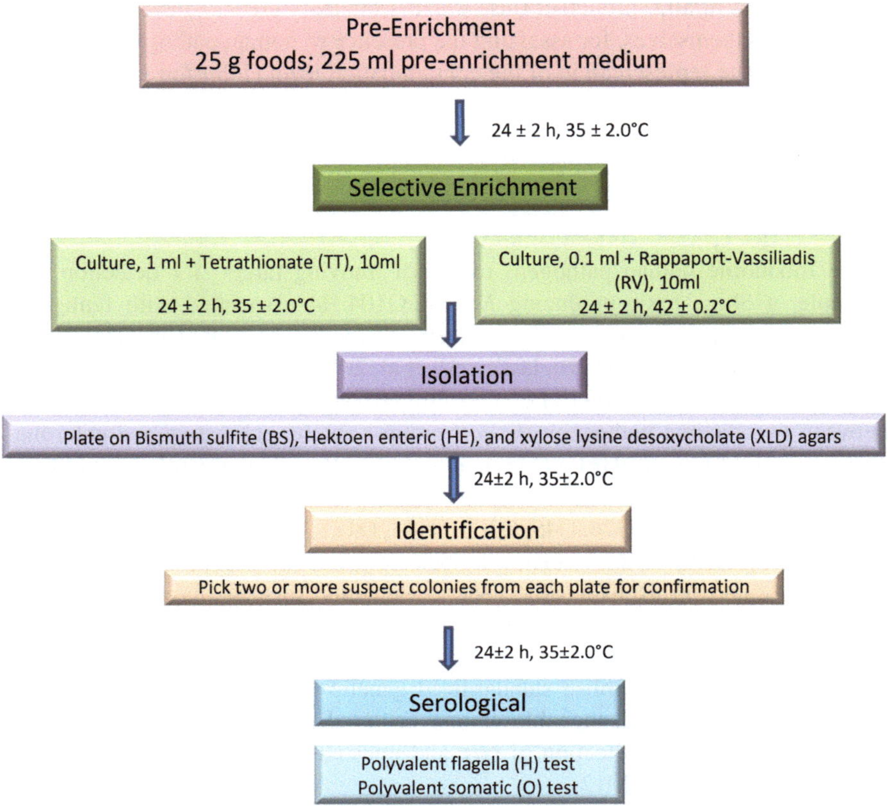

Fig. 2 Current protocol of the Food and Drug Administration for detection of *Salmonella* in food

Agriculture (USDA). Both agencies have developed and validated sets of foodborne human pathogen detection methods in various foods; these are published online as FDA's Bacteriological Analytical Manual (BAM) (U.S. FDA 2013a) and USDA's Microbiology Laboratory Guidebook (MLG) (USDA 2013), respectively. The methods presented in these two manuals are the current standards for routine outbreak investigations and regulatory assays. USDA has primary responsibility for the safety of meat, poultry, and certain egg products, while FDA regulates all the remaining foods that are not regulated by USDA, including foods of plant origin; therefore, BAM protocols are considered to be the standard assays for HPOP detection.

In general, the standard assays consist of pre-enrichment, selective enrichment, plating on differential media, and biochemical, serological, or molecular tests for strain identification. A schematic diagram of the current standard assay for detection of *Salmonella* on plant products is shown in Fig. 2. The pre-enrichment media are not selective and encourage the growth of microbes in general. Depending on the commodity being tested, trypticase soy broth (TSB), buffered peptone water (BPW), universal pre-enrichment broth (UPB), or lactose broth (LB) are specified as the pre-enrichment broths in the standard assays. Tetrathionate broth (TT) and Rappaport-Vassiliadis broth (RV) are used as selective enrichment media. Bismuth sulfite (BS), Hektoen enteric (HE), and xylose lysine desoxycholate (XLD) agars are the standard selective solid media for *Salmonella*. As alternatives to the conventional biochemical tests, commercial biochemical utilization assay systems such as API 20E, Enterotube II, *Enterobacteriaceae* II, MICRO-ID, or Vitek GNI, are used for presumptive generic identification of foodborne *Salmonella*. The standard assay, which consists of multiple incubation steps, is slow, requiring 4–5 days. Molecular-based methods, offering improved sensitivity, specificity, and speed, have been developed and some have been well-validated and incorporated into the standard assays or are being introduced as alternatives. For example, enriched food samples can be screened by real-time PCR to rule out negative samples and establish the presumptive presence of *E. coli* O157:H7 within 24 h (BAM Chapter 4A, U.S. FDA 2013a).

Molecular-Based Methods

Many molecular-based methods have been developed in the pursuit for rapid, reliable, sensitive, and robust methods for the detection of foodborne human pathogens. These include polymerase chain reaction (PCR)-based methods, microarray, next-generation sequencing and biosensor techniques.

PCR-Based Methods. Polymerase chain reaction (PCR) has become an important diagnostic tool for HPOPs and many PCR-based detection methods have been used routinely for sensitive and specific screening of foodborne pathogens and for isolate confirmation tests (Mothershed and Whitney 2006).

Conventional and Multiplex PCR. Since the early 1990s various genes, including those for 16S rRNA and virulence, have been targeted for the detection of human pathogens in conventional and multiplex PCR for routine rapid screening tests as well as for confirmation of isolate identity. Over 30 genes have been used in the detection of *Salmonella*, with *invA* (encoding invasin A) being targeted most frequently. Differentiation of *E. coli* O157:H7 from other pathogenic *E. coli* strains, as well as generic *E. coli*, requires the use of multiple genes in a multiplex PCR format. The targeted genes usually include *stx1 and stx2* (encoding Shiga toxin 1 and 2) and +93 *uidA* (encoding –D-glucuronidase with +93 single nucleotide polymorphisms) for screening and *stx1*, *stx2*, +93*uidA*, γ-eaeA (encoding intimin), and *ehxA* (encoding enterohemolysin), or *rfbE* (encoding O-antigen production specific to O157) for isolate confirmation (U.S. FDA 2013a; Paton and Paton 1998; Abdulmawjood et al. 2003). The gene most often targeted for the detection of *L. monocytogenes*, *hly*, encodes the pore-forming cytolysin listeriolysin

(Aznar and Alarcón 2002; Brehm-Stecher and Johnson 2007). The BAX® system (DuPont Qualicon) was one of the first commercial PCR-based automated systems for the detection of human pathogens, including *Salmonella*, *E. coli* O157:H7, and *L. monocytogenes* (Shearer et al. 2001).

Simultaneous detection of several human pathogens in a single multiplex PCR has gained popularity as it is more cost effective and time saving than conventional PCR. Fratamico and Strobaugh (1998) developed a multiplex PCR targeting the *eaeA*, *stx1*, *stx2*, and *invA* genes for the detection of *E. coli* O157:H7 and *Salmonella* in apple cider. A multiplex PCR targeting *E. coli* O157:H7 (+93 *uid* A), *Salmonella* (unknown gene), and *Shigella* (*ipa*), developed by Li and Mustapha (2004) and applied to apple cider and fresh produce, was able to detect the three pathogens at initial inoculation levels of 0.8 CFU/g (or ml) in cider, cantaloupe, lettuce, tomato, and watermelon and of 80 CFU/g in alfalfa, following 34 h enrichment. Recently, Timmons et al. (2012) developed a multiplex PCR for the detection of *Salmonella* and STEC in leafy greens and tomatoes with 10–100-fold higher sensitivity when the primers were modified by the addition of a 5′-flap.

Real Time (Quantitative) PCR. With added specificity, sensitivity, and speed, real time PCR has become very popular in recent years for the detection of human pathogens. When Liming et al. (2004) compared two commercially available PCR kits for the detection of *L. monocytogenes* in fresh produce, a conventional PCR (BAX, DuPont Qualicon) and a molecular beacon-based real-time PCR (iQ-Check, Bio-Rad Laboratories), as few as 4–7 CFU of *L. monocytogenes*/25 g produce could be detected by both kits after enrichment. However, the real-time PCR assay required considerably shorter time than the conventional format (26 h versus 52 h), and offered increased specificity. TaqMan-based real-time PCR assays, targeting the genes *invA* (*Salmonella*) and *hly* (*L. monocytogenes*), coupled with International Standard Organization (ISO) enrichment, were evaluated for detection of *L. monocytogenes* and *Salmonella* in fresh fruit and vegetables including tomato, lettuce, apple, grape, watercress, and soybean sprouts (Badosa et al. 2009). Both pathogens were detected consistently in all samples except soybean sprouts, with a limit of detection (LOD) of 1 CFU/25 g sample, after enrichment. The LOD for soybean sprouts ranged from 1 to 10 CFU/25 g sample after enrichment. The standard FDA assays for the detection of *E. coli* O157:H7 include the use of real-time PCR in sample screening and isolate confirmation as described above. A number of commercial real-time PCR based systems for foodborne human pathogen detection are available, including the BAX® system (DuPont Qualicon), TaqMan® and MicroSEQ® food pathogen detection kits (Life Technologies), iQ Check (Bio-Rad), and foodproof® (Merck).

Similar to conventional PCR, multiplex real-time PCR has been applied for simultaneous detection of several major human pathogens in produce (Bhagwat 2003); the method was able to detect 1–10 cells/ml of *E. coli* O157 and *Salmonella*, and 100–1,000 cells/ml of *L. monocytogenes*, after enrichment. The recent emergence of more than 100 non-O157 STEC strains associated with human illness and the increased frequency of non-O157 STEC isolates implicated in hemolytic-uremic syndrome led the USDA to expand the zero-tolerance policy for *E. coli* O157:H7 to include the top six non-O157 STEC serogroups in raw and nonintact beef products (USDA 2013). Consequently, multiplex real-time PCR assays targeting the top seven STECs (O157, O26, O45, O103, O111, O121, and O145), have been developed (Anklam et al. 2012). The assays, which consist of four sets of multiplex PCRs, target the O-antigen gene clusters for each serogroup, *uidA* for generic *E. coli* (also serves as internal amplification control), *rfbE* for O157, and several virulence genes (*stx1*, *stx2*, *eae*, and *ehxA*). The LODs of the assays were 10^3 or 10^4 CFU/ml for pure culture or spiked fecal samples, respectively, and 1 CFU/ml after 6 h enrichment.

One concern with the detection of human pathogens by PCR is its inability to distinguish whether the DNA is from dead or living cells, could result in overestimation of contamination levels in food when it is used as a screening test. Propidium monazide (PMA), a DNA intercalating reagent, penetrates only into cells having compromised cell membranes (nonviable cells), cross-linking their DNA and thereby rendering them inaccessible to downstream PCR amplification (Nocker et al. 2006). Sample pretreatment with PMA prior to real-time PCR has been applied in the detection of viable human pathogens on plant products (Dinu and Bach 2013; Elizaquível et al. 2012, 2013). PMA treatment becomes especially critical in using real-time PCR as a quantitative assay (qPCR) to evaluate the efficacy of various disinfection technologies, in which a large number of target human pathogen cells are dead from disinfection treatment but their DNAs still are detected by qPCR unless there has been a prior PMA treatment (Elizaquível et al. 2012, 2013).

Messenger RNAs, which most likely associate only with live cells as they have much shorter half-lives than DNA, have been targeted in real-time PCR (termed real-time reverse transcriptase PCR or rt-RT-PCR) for the detection of viable human pathogens (Dupray et al. 1997). Miller et al. (2010, 2011) developed an rt-RT-PCR (targeting *invA* mRNA) for the detection of *Salmonella enterica*; when tested on fresh produce, including lettuce, tomatoes, Jalapeno and Serrano peppers, the detection limits were ca. 400 CFU/g with 6 h enrichment and ca. 4×10^5 CFU/g without enrichment.

Another strategy in eliminating false-positive results from nonviable pathogen cells and improving real-time PCR assay sensitivity is the addition of an enrichment step in the detection of human pathogens. Krascsenicsova et al (2008) developed two-step enrichments consisting of overnight enrichment in buffered peptone water followed by a 5 h subculture in Rappaport-Vassliadis medium, and a *Salmonella*-specific real-time PCR screening of the bacterial cell lysates from RV subculture. Following this protocol, the detection limit for dead *Salmonella* cells in artificially contaminated foods was 10^7 dead cells/25 g food, which significantly exceeds the dead cell concentration that could be encountered normally in a food system, therefore eliminating potential false positive results. When compared with the standard microbiological method EN ISO 6579 with 36 food samples artificially contaminated with *Salmonella* at a level of 1 CFU/25 g food, identical results were obtained from both methods. Chen et al (2011) developed a real time PCR assay targeting the *iap* gene for the detection of *L. monocytogenes* in selected food matrices, including alfalfa sprouts. Without enrichment, the detection limit was 4×10^4 CFU/ml broth (corresponding to 4×10^5 CFU/g food), however, with 24 h enrichment the detection limit improved to 9 CFU/25 g alfalfa sprouts. Similar approaches have been applied for other human pathogen detections in plant products (O'Grady et al. 2008, 2009; Kotzekidou 2013; Weagant et al. 2011). In general, combining a short enrichment with molecular-based detection can improve the assay's sensitivity, reduce false-positive results due to DNA from nonviable pathogen cells, and shorten the assay time from 4–5 days to 1 day.

Loop Mediated Isothermal Amplification (LAMP). LAMP, unlike PCR, involves an auto-cyclic isothermal amplification (60–65 °C) of target DNA by strand-displacing *Bst* DNA polymerase (a large fragment of *Bacillus stearothermophilus* DNA polymerase), which can be accomplished in a water bath or heating block, eliminating the need for a thermocycler (Notomi et al. 2000; Mori and Notomi 2009). The final products can be visualized as a turbidity or color change in the reaction tube, therefore saving time compared to the gel electrophoresis required for PCR. Hence, LAMP has become a promising technology for the rapid detection of human pathogens in the field and has been used widely as a rapid, specific, sensitive, and cost-effective pathogen detection method. Wang et al. (2012) developed a LAMP assay targeting *E. coli* O157 and six other top serotypes of Shiga toxin-producing *E. coli* (*stx1, stx2, eae, wzx* or *wzy* genes) and tested it in produce samples. The assay was rapid (40 min), specific, and sensitive (ca. 1–20 CFU/reaction), with an improved sensitivity (1–2 CFU/25 g sample) when combined with a brief enrichment step (6–8 h). Commercial test kits based on LAMP for the detection of *E. coli* O157:H7, *Salmonella*, and *Listeria* spp. are available (3 M Molecular Detection System). With the test kits, the amplification and detection processes can be completed within 75 min with positive results available as early as 15 min. However, an overnight single enrichment step is still required for detection of low numbers of human pathogens in plant products.

DNA Microarrays. Even though multiplex PCR is able to detect multiple targets in a single reaction, the detection capacity is still restricted to a few targets per assay because of the limited choices of fluorescent dyes (Gannon et al. 1997; Yamazaki et al. 2007). DNA microarrays, also referred to as a lab-on-a-chip technology, are promising high-throughput tools for multiple foodborne human pathogen detection. A microarray consists of a solid matrix, usually a glass slide, on which oligonucleotide probes or other DNA fragments are placed in very precise locations at high density. Target DNA sequences in a sample are then hybridized to the probes and detected by fluorescence. The advantage of microarray-based detection is the combined powerful nucleic acid amplification strategies with a massive screening capability, resulting in a high level of sensitivity, specificity, and high throughput capacity; it can detect many different human pathogens in a single assay. While numerous microarrays have been described for the detection of a wide range of foodborne human pathogens, the performance and cost remain limiting factors for routine detection in food samples. With advances in fabrication, robotics, and bioinformatics, microarray technology could be improved in terms of efficiency, discriminatory power, reproducibility, sensitivity, and specificity (Volokhov et al. 2002; Jin et al. 2005; Suo et al. 2010; Severgnini et al. 2011).

Next-Generation Sequencing. Next-generation sequencing (NGS) technology, also known as pyrosequencing or high-throughput sequencing, is revolutionizing not only the study of microbial ecology but also the detection of human pathogens in a variety of samples. Unlike other molecular-based detection methods, which require prior knowledge of sequence information on the pathogens to be detected, the NGS approach is unlimited, making it possible to detect any known and novel pathogens in a single assay (Amoako 2013; Eom et al. 2007).

Although NGS has facilitated remarkable advances in the study of microbial ecology in the past few years, the most critical challenge for its application in pathogen detection is the development of improved, user-friendly bioinformatics and visualization platforms to facilitate rapid and robust analysis and interpretation of high volume of sequencing data. Our institute has developed a bioinformatic process, termed E-probe Diagnostic Nucleic acids Analysis (EDNA), a series of bioinformatics pipelines in which e-probes are generated and validated using a mock sample database, then used to detect plant and foodborne human pathogens in complex samples from metagenomic sequencing data (Stobbe et al. 2013). EDNA eliminates the need for assembly and GenBank BLAST steps while finding nucleic acid signatures of microbes of interest. *In silico* simulations indicated that the procedure was both sensitive and specific in the detection of RNA and DNA viruses, as well as both prokaryotic and eukaryotic organisms. EDNA specificity may be even better when it is applied to actual infected plant samples than in *in silico* simulations.

Biosensor-Based Techniques

Recently, many researchers have turned to the development of biosensor-based detection techniques. Since these approaches are amenable to miniaturization with minimum sample preparation they could be used as on-site or in-field detection tools. A typical biosensor consists of a biological probe that captures the target, a transduction platform that generates a measurable signal in the event of target capture, and an amplifier which processes the signal to give a measurement (Velusamy et al. 2009; Arora et al. 2011; Yoon and Kim. 2012; Singh et al. 2013). Biological probes employed on biosensors include nucleic acids, antibodies, bacterial phages (genetically modified phage display peptides), and phage receptor binding proteins. Signal amplification and detection can be optical, or can utilize sensors based on surface plasmon resonance, bioluminescence, fluorescence, quartz crystal macrobalance, or magnetoelastics. For example, Park et al. (2013) evaluated a phage-based magnetoelastic (ME) biosensor for direct detection of *Salmonella* Typhimurium on spinach leaves. The limit of detection (LOD) for the ME biosensor was 2.17 and 1.94 log CFU/spinach leaf for adaxial and abaxial surfaces, respectively, compared to 1.37 log CFU/spinach leaf obtainable by real-time quantitative PCR. Tlili et al. (2013) reported bacteria screening, viability, and confirmation assays using bacteriophage-impedimetric/loop-mediated isothermal amplification dual-response biosensors. They suggested that integration of the T4-bacteriophage-modified biosensor and LAMP can achieve screening, viability, and confirmation in less than 1 h. Biosensors for detecting food samples have not yet been commercialized, as new bio-molecular techniques are needed to improve biosensor characteristics such as sensitivity, specificity, and easy integration onto the transduction platform. Nevertheless, biosensors offer real time detection, rather than requiring hours or days as the assays described above.

Conclusions

With the rapid increase of international trade, greater public consumption of raw produce as a healthy eating habit, and more out-of-home dining, disease outbreaks attributable to human pathogens on plant products are becoming more common and widespread. Scientific information related to the survival, growth and distribution of human pathogens on plants and in the environments where plant products are grown, processed, transported, stored, and consumed are lacking. In a rapidly changing world, new surveillance data on these pathogens are urgently needed. Effective, accurate, specific and rapid detection methods are essential for surveillance, outbreak investigation, and enforcement of government food safety regulations. In most cases, traditional culture detection and isolation methods remain the 'gold standards'. New enrichment broths and media (especially pathogen-selective media) have made culture detection methods more sensitive and accurate in recent years; these areas will continue to be a focus. At the same time, however, molecular methods such as PCR, LAMP, DNA microarrays, next generation sequencing, and biosensors, are being explored with increasing enthusiasm. Currently, although PCR is the most widely used and most successful molecular method for detection of foodborne pathogens its specificity, accuracy, and repeatability do not yet consistently match or exceed those of the traditional methods. New variations, such as loop mediated isothermal amplification also have great potential. DNA sequencing and whole genome sequencing also have been significantly improved for applications in food safety. However, a major drawback of all molecular detection methods remains: they cannot produce a live microbe isolate for use in government regulatory functions and future scientific studies. Effective detection and diagnostic methods of the future will continue to differ in features depending upon the intended application and operators.

References

Abdulmawjood A, Bültea M, Cook N, Rotha S, Schönenbrüchera H, Hoorfar J (2003) Toward an international standard for PCR-based detection of *Escherichia coli* O157. Part 1. Assay development and multi-center validation. J Microbiol Methods 55:775–786

Amoako KK (2013) Rapid detection and identification of *Bacillus anthracis* in food using pyrosequencing technology. Int J Food Microbiol 165:319–325

Anklam KS, Kanankege KS, Gonzales TK, Kaspar CW, Döpfer D (2012) Rapid and reliable detection of Shiga toxin-producing *Escherichia coli* by real-time multiplex PCR. J Food Prot 75:643–650

Arora P, Sindhu A, Dilbaghi N, Chaudhury A (2011) Biosensors as innovative tools for the detection of food borne pathogens. Biosens Bioelectron 28:1–12

Aznar R, Alarcón B (2002) On the specificity of PCR detection of *Listeria monocytogenes* in food: a comparison of published primers. System Appl Microbiol 25:109–119

Badosa E, Chico N, Pla M, Parés D, Montesinos E (2009) Evaluation of ISO enrichment real-time PCR methods with internal amplification control for detection of *Listeria monocytogenes* and *Salmonella enterica* in fresh fruit and vegetables. Lett Appl Microbiol 49:105–111

Barak JD, Schroeder BK (2012) Interrelationships of food safety and plant pathology: the life cycle of human pathogens on plants. Annu Rev Phytopathol 50:241–266. doi:10.1146/annurev-phyto-081211-172936

Behravesh CB, Mody RK, Jungk J, Gaul L, Redd JT, Chen S, Cosgrove S, Hedican E, Sweat D, Chavez-Hauser L, Snow SL, Hanson H, Nguyen TA, Sodha SV, Boore AL, Russo E, Mikoleit M, Theobald L, Gerner-Smidt P, Hoekstra RM, Angulo FJ, Swerdlow DL, Tauxe RV, Griffin PM, Williams IT (2011) 2008 outbreak of *Salmonella Saintpaul* infections associated with raw produce. N Engl J Med 364:918–927. doi:10.1056/NEJMoa1005741

Bhagwat AA (2003) Simultaneous detection of *Escherichia coli* O157:H7, Listeria monocytogenes and Salmonella strains by real-time PCR. Int J Food Microbiol 84:217–224

Bielaszewska M, Mellmann A, Zhang W, Kock R, Fruth A, Bauwens A, Peters G, Karch H (2011) Characterization of the *Escherichia coli* strain associated with an outbreak of hemolytic-uremic syndrome in Germany, 2011: a microbiological study. Lancet Infect Dis 11:671–676

Bloch SK, Felczykowska A, Nejman-Falenczyk B (2012) *Escherichia coli* O104:H4 outbreak – have we learned a lesson from it? Acta Biochim Pol 59:483–484

Bowen A, Fry A, Richards G, Beuchat L (2006) Infections associated with cantaloupe consumption: a public health concern. Epidemiol Infect 134:675–685. doi:10.1017/S0950268805005480

Brandl MT (2006) Fitness of human enteric pathogens on plants and implications for food safety. Annu Rev Phytopathol 44:367–392. doi:10.1146/annurev.phyto.44.070505.143359

Brehm-Stecher BF, Johnson EA (2007) Rapid detection of *Listeria*. In: Marth E, Ryser E (eds) *Listeria*, listeriosis and food safety, 3rd edn. Marcel Dekker, New York

CDC (2011) Multistate outbreak of Listeriosis linked to whole cantaloupes from Jensen Farms, Colorado. http://www.cdc.gov/listeria/outbreaks/cantaloupes-jensen-farms/index.html. Accessed 21 Aug 2012

Chen Y, Kumar N, Sidddique N (2011) Development and evaluation of a real-time polymerase chain reaction assay targeting iap for the detection of *Listeria monocytogenes* in select food matrices. Foodborne Pathog Dis 8:1063–1069

Danhorn T, Fuqua C (2007) Biofilm formation by plant-associated bacteria. Annu Rev Microbiol 61:401–422. doi:10.1146/annurev.micro.61.080706.093316

Delibato E, Anniballi F, Vallebona PS, Palleschi G, Volpe G, Losio MN, De Medici D (2013) Validation of a 1-day analytical diagnostic real-time PCR for the detection of *Salmonella* in different food meat categories. Food Anal Methods 6:996–1003

Dinu LD, Bach S (2013) Detection of viable but non-culturable *Escherichia coli* O157:H7 from vegetable samples using quantitative PCR with propidium monoazide and immunological assays. Food Control 31:268–273

Dupray E, Caprais MP, Derrien A, Fach P (1997) *Salmonella* DNA persistence in natural seawaters using PCR analysis. J Appl Bacteriol 82:507–510

Elizaquível P, Sánchez G, Aznar R (2012) Application of propidium monoazide quantitative PCR for selective detection of live *Escherichia coli* O157:H7 in vegetables after inactivation by essential oils. Int J Food Microbiol 159:115–121

Elizaquível P, Maryam A, Gloria S, Rosa A (2013) Evaluation of *Zataria multiflora Boiss.* essential oil activity against *Escherichia coli* O157:H7, *Salmonella enterica* and *Listeria monocytogenes* by propidium monoazide quantitative PCR in vegetables. Food Control 34:770–776

Eom HS, Hwang BH, Kim DH, Lee IB, Kim YH, Cha HJ (2007) Multiple detection of foodborne pathogenic bacteria using a novel 16S rDNA-based oligonucleotide signature chip. Biosens Bioelectr 22:845–853

Fletcher J, Leach J, Eversole K, Tauxe R (2013) Human pathogens on plants: designing a multidisciplinary strategy for research. Phytopathology 103:306–315

Fratamico P, Strobaugh TP (1998) Simultaneous detection of *Salmonella* spp. and *Escherichia coli* O157:H7 by multiplex PCR. J Ind Microbiol Biotechnol 21:92–99

Gannon VPJ, D'Souza S, Graham T, King RK, Rahn K, Read S (1997) Use of the flagellar H7 gene as a target in multiplex PCR assays and improved specificity in identification of enterohemorrhagic *Escherichia coli* strains. J Clin Microbiol 35:656–662

Gupta SK, Nalluswami K, Snider C, Perch M, Balasegaram M, Burmeister D, Lockett J, Sandt C, Hoekstra RM, Montgomery S (2007) Outbreak of *Salmonella Braenderup* infections associated with Roma tomatoes, northeastern United States, 2004: a useful method for subtyping exposures in field investigations. Epidemiol Infect 135:1165–1173

Hilborn ED, Mermin JH, Mshar PA, Hadler JL, Voetsch A, Wojtkunski C, Swartz M, Mshar R, Lambert-Fair MA, Farrar JA, Glynn MK, Slutsker L (1999) A multistate outbreak of *Escherichia coli* O157:H7 infections associated with consumption of mesclun lettuce. Arch Int Med 159:1758–1764

Jin HY, Tao KH, Li YX, Li FQ, Li SQ (2005) Microarray analysis of *Escherichia coli* O157:H7. World J Gastroenterol 11:5811–5815

Johnston LM, Elhanafi D, Drake M, Jaykus L (2005) A simple method for the direct detection of *Salmonella* and *Escherichia coli* O157:H7 from raw alfalfa sprouts and spent irrigation water using PCR. J Food Prot 68:2256–2263

Kotzekidou P (2013) Survey of *Listeria monocytogenes*, *Salmonella* spp. and *Escherichia coli* O157:H7 in raw ingredients and ready-to-eat products by commercial real-time PCR kits. Food Microbiol 35:86–91

Krascsenicsova K, Piknova L, Kaclikova E, Kuchta T (2008) Detection of *Salmonella enterica* in food using two-step enrichment and real-time polymerase chain reaction. Lett Appl Microbiol 46:483–487

Li Y, Mustapha A (2004) Simultaneous detection of *Escherichia coli* O157:H7, *Salmonella*, and *Shigella* in apple cider and produce by a multiplex PCR. J Food Prot 67:27–33

Liming SH, Zhang Y, Meng J, Bhagwat AA (2004) Detection of *Listeria monocytogenes* in fresh produce using molecular beacon – real-time PCR technology. J Food Sci 69:M240–M245

Lynch MF, Tauxe RV, Hedberg CW (2009) The growing burden of foodborne outbreaks due to contaminated fresh produce: risks and opportunities. Epidemiol Infect 137:307–315

Mellmann A, Harmsen D, Cummings CA, Zentz EB, Leopold SR, Rico A, Prior K, Szczepanowski R, Ji Y, Zhang W, McLaughlin SF, Henkhaus JK, Leopold B, Bielaszewska M, Prager R, Brzoska PM, Moore RL, Guenther S, Rothberg JM, Karch H (2011) Prospective genomic characterization of the German enterohemorrhagic *Escherichia coli* O104:H4 outbreak by rapid next generation sequencing technology. PLoS One 6:e22751. doi:10.1371/journal.pone.002275

Miller ND, Draughon FA, D'Souza DH (2010) Real-time reverse-transcriptase-polymerase chain reaction for *Salmonella enterica* detection from jalapeño and serrano peppers. Foodborne Pathog Dis 7:367–373

Miller ND, Davidson PM, D'Souza DH (2011) Real-time reverse-transcriptase PCR for *Salmonella* Typhimurium detection from lettuce and tomatoes. Food Sci Tech 44:1088–1097

Mody RK, Greene SA, Gaul L, Sever A, Pichette S, Zambrana I, Dang T, Gass A, Wood R, Herman K, Cantwell LB, Falkenhorst G, Wannemuehler K, Hoekstra RM, McCullum I, Cone A, Franklin L, Austin J, Delea K, Behravesh CB, Sodha SV, Yee JC, Emanuel B, Al-Khaldi SF, Jefferson V, Williams IT, Griffin PM, Swerdlow DL (2011) National outbreak of *Salmonella* serotype saintpaul infections: importance of Texas restaurant investigations in implicating jalapeno peppers. PLoS One 6:e16579. doi:10.1371/journal.pone.0016579

Mori Y, Notomi T (2009) Loop-mediated isothermal amplification (LAMP): a rapid, accurate, and cost-effective diagnostic method for infectious diseases. J Infect Chemother 15:62–69

Mothershed EA, Whitney AM (2006) Nucleic acid-based methods for the detection of bacterial pathogens: present and future considerations for the clinical laboratory. Clin Chim Acta 363:206–220

Nocker A, Cheung CY, Camper AK (2006) Comparison of propidium monoazide with ethidium monoazide for differentiation of live vs. dead bacteria by selective removal of DNA from the cells. J Microbiol Met 67:310–320

Notomi T, Okayama H, Masubuchi H, Yonekawa T, Watanabe K, Amino N, Hase T (2000) Loop-mediated isothermal amplification of DNA. Nucleic Acids Res 28:E63

O'Grady J, Sedano-Balbás S, Maher M, Smith TJ, Barry T (2008) Rapid real-time PCR detection of *Listeria monocytogenes* in enriched food samples based on the ssrA gene, a novel diagnostic target. Food Microbiol 25:75–84

O'Grady J, Ruttledge M, Sedano-Balbás S, Smith TJ, Barry T, Maher M (2009) Rapid detection of *Listeria monocytogenes* in food using culture enrichment combined with real-time PCR. Food Microbiol 26:4–7

Park MK, Weerakoon KA, Oh JH, Chin BA (2013) The analytical comparison of phage-based magnetoelastic biosensor with Taq Man-based quantitative PCR method to detect *Salmonella* Typhimurium on cantaloupes. Food Control 33:330–336

Paton AW, Paton JC (1998) Detection and characterization of Shiga toxigenic *Escherichia coli* by using multiplex PCR assays for stx_1, stx_2, *eaeA*, enterohemorrhagic *E. coli hlyA*, rfb_{O111}, and rfb_{O157}. J Clin Microbiol 36:598–602

Severgnini M, Cremonesi P, Consolandi C, de Bellis G, Castiglioni B (2011) Advances in DNA microarray technology for the detection of foodborne pathogens. Food Bioprocess Technol 4:936–953

Shearer AEH, Strapp CM, Joerfer RD (2001) Evaluation of a polymerase chain reaction-based system for detection of *Salmonella enteritidis*, *Escherichia coli* O157:H7, *Listeria* spp, and *Listeria monocytogenes* on fresh fruits and vegetables. J Food Prot 64:788–795

Singh A, Poshtiban S, Evoy S (2013) Recent advances in bacteriophage based biosensors for food-borne pathogen detection. Sensors 13:1763–1786

Sivapalasingam S, Friedman CR, Cohen L, Tauxe RV (2004) Fresh produce: a growing cause of outbreaks of foodborne illness in the United States, 1973 through 1997. J Food Prot 67:2342–2353

Stobbe T, Daniels J, Espindola A, Melcher U, Ochoa Corona F, Garzon C, Verma R, Fletcher J, Schneider W (2013) Electronic diagnostic nucleic acid analysis (EDNA): a theoretical approach for improved handling of massively parallel sequencing data for diagnostics. J Micr Meth. http://dx.doi.org/10.1016/j.mimet.2013.07.002

Suo B, He Y, Paoli G, Gehring A, Tu SI, Shi X (2010) Development of an oligonucleotide-based microarray to detect multiple foodborne pathogens. Mol Cell Probes 24:77–86

Teplitski M, Barak JD, Schneider KR (2009) Human enteric pathogens in produce: un-answered ecological questions with direct implications for food safety. Curr Opin Plant Biotechnol 20:166–171

Timmons C, Dobhal S, Fletcher J, Ma LM (2012) Primers with 5′ flaps improve the efficiency and sensitivity of multiplex PCR assays for the detection of *Salmonella* and *Escherichia coli* O157:H7. J Food Prot 76:668–673

Tlili C, Sokullu E, Safavieh M, Tolba M, Ahmed MU, Zourob M (2013) Bacteria screening, viability, and confirmation assays using bacteriophage-impedimetric/loop-mediated isothermal amplification dual-response biosensors. Analyt Chem 85:4893–4901

Twardoń J, Sobieszczańska B, Gonet A, Błaszkowska M (2005) Epidemiology of Shiga-like toxin-producing *Escherichia coli* strains (STEC). Elect J Polish Agric Univ 8, #03. Online: http://www.ejpau.media.pl/volume8/issue4/art-03.html. Accessed 8 Oct 2013

U.S. Department of Agriculture (2013) Microbiology laboratory guidebook. http://www.fsis.usda.gov/wps/portal/fsis/topics/science/laboratories-and-procedures/guidebooks-and-methods/microbiology-laboratory-guidebook/microbiology-laboratory-guidebook. Accessed 8 Nov 2013

U.S. Food & Drug Administration (2013a) Bacteriological Analytical Manual (BAM). http://www.fda.gov/food/foodscienceresearch/laboratorymethods/ucm2006949.htm. Accessed 8 Nov 2013

U.S. Food & Drug Administration (2013b) The reportable food registry third annual report: targeting inspection resources and identifying patterns of adulteration. http://www.fda.gov/downloads/Food/ComplianceEnforcement/RFR/UCM349856.pdf. Accessed 8 Nov 2013

Velusamy V, Arsha K, Korostynska O, Oliwa K, Adley C (2009) An overview of foodborne pathogen detection: in the perspective of biosensors. Biotechnol Adv 28:232–254

Vidaver AK, Tolin S, Lambrecht P (2006) Laboratory, growth chamber and greenhouse microbial safety: plant pathogens and plant associated microorganisms of significance to human health. In: Fleming DO, Hunt DL (eds) Biological safety: principles and practices. ASM Press, Washington, DC

Volokhov D, Rasooly A, Chumakov K, Chizhikov V (2002) Identification of *Listeria* species by microarray-based assay. J Clin Microbiol 40:4720–4728

Wang F, Jiang L, Yang Q, Prinyawiwatkul W, Ge B (2012) Rapid and specific detection of Escherichia coli serogroups O26, O45, O103, O111, O121, O145, and O157 in ground beef, beef trim, and produce by loop-mediated isothermal amplification. Appl Environ Microbiol 78:2727–2736

Warren BR, Yuk H-G, Schneider KR (2007) Detection of *Salmonella* by flow-through immunocapture real-time PCR in selected foods within 8 hours. J Food Prot 70:1002–1006

Weagant SD, Jinneman KC, Yoshitomi KJ, Zapata R, Fedio WM (2011) Optimization and evaluation of a modified enrichment procedure combined with immunomagnetic separation for detection of *E. coli* O157:H7 from artificially contaminated alfalfa sprouts. Int J Food Microbiol 149:209–217

Wendel AM, Johnson DH, Sharapov U, Grant J, Archer JR, Monson T, Koschmann C, Davis JP (2009) Multistate outbreak of *Escherichia coli* O157:H7 infection associated with consumption of packaged spinach, August-September 2006: the Wisconsin investigation. Clin Infect Dis 48:1079–1086. doi:10.1086/597399

Wheeler C, Vogt TM, Armstrong GL, Vaughan G, Weltman A, Nainan OV, Dato V, Xia G, Waller K, Amon J, Lee TM, Highbaugh-Battle A, Hembree C, Evenson S, Ruta MA, Williams IT, Fiore AE, Bell BP (2005) An outbreak of hepatitis A associated with green onions. N Engl J Med 353:890–897. doi:10.1056/NEJMoa050855

Yamazaki-Matsune W, Taguchi M, Seto K, Kawahara R, Kawatsu K, Kumeda Y, Nukina M, Misawa N, Tsukamoto T (2007) Development of a multiplex PCR assay for identification of *Campylobacter coli*, *Campylobacter fetus*, *Campylobacter hyointestinalis* subsp. *hyointestinalis*, *Campylobacter jejuni*, *Campylobacter lari* and *Campylobacter upsaliensis*. J Med Microbiol 56:1467–1473

Yoon J, Kim B (2012) Lab-on-a-Chip pathogen sensors for food safety. Sensors 12:10713–10741

Yoshitomi KJ, Jinneman KC, Zapata R, Weagant SD, Fedio WM (2012) Detection and isolation of low levels of *E. coli* O157:H7 in cilantro by real-time PCR, immunomagnetic separation, and cultural methods with and without an acid treatment. J Food Sci 77:M481–M489

Plant Disease Diagnostics for Forensic Applications

Jacqueline Fletcher, Francisco M. Ochoa Corona, and Mark Payton

Abstract Although most plant diseases are the result of natural or unintentional causes, cropping systems and essential natural plant resources such as forests and grasslands also are considered vulnerable to actions of nefarious intent. Microbial forensics is defined as the application of scientific approaches to solving a crime that involves a microorganism; its goal is to investigate and present unbiased scientific evidence useful for attributing the crime to a perpetrator. Recent programs intended to enhance general capabilities in microbial forensics have included specific attention to plant pathogens. Compared to the strategies employed by traditional plant disease diagnosticians, forensic applications of plant pathogen diagnostics require unusually high levels of stringency, reliability, and prior validation. These assays must be paired with court-defensible sampling methods, chain of custody, and other traditional and non-traditional methods of forensic investigation.

Keywords Forensic science • Microbial forensics • Plant pathogen forensics • Biosecurity • Disease diagnosis • Sampling • Chain of custody • Pathogen detection • Pathogen discrimination • Validation • Sensitivity • Specificity • Reproducibility • Exclusivity • Inclusivity

J. Fletcher (✉) • F.M. Ochoa Corona
Department of Entomology & Plant Pathology and National Institute for Microbial Forensics & Food and Agricultural Biosecurity, Stillwater, OK 74078, USA
e-mail: jacqueline.fletcher@okstate.edu

M. Payton
Department of Statistics, Oklahoma State University, Stillwater, OK 74078, USA

M.L. Gullino and P.J.M. Bonants (eds.), *Detection and Diagnostics of Plant Pathogens*, Plant Pathology in the 21st Century 5, DOI 10.1007/978-94-017-9020-8_7,

Forensics and Plant Pathology – Synergy of Two Disciplines

Forensic plant pathology is a blend of the disciplines of plant pathology and forensic science that supports the investigation of plant diseases and pathogens by providing unbiased scientific methodology and evidence for criminal attribution. Important to this effort are traceback strategies for determining pathogen origin and movement pathway(s) as well as the possible role of human intent. Plant pathology and forensic science find a common arena with other disciplines within the realm of agricultural biosecurity, which includes science-based policies, measures and regulatory frameworks for reacting to and managing risks associated with food, agriculture, forestry, and the environmental (UN 2002). The concept also includes other areas of environmental risk such as aquatic systems, and strongly associated sectors such as human health, justice and defense (Ochoa-Corona 2011). Unwanted movement of plant pathogens and pests into the agricultural sector can take place by various means including wind, water, insects, international commerce and travel. Such movement occurs frequently as inadvertent introduction of exotic pathogens and pests; however smuggling and illegal trade are examples of relevant criminal activities. In such situations multidisciplinary teams including representatives of the diagnostic, regulatory, and law enforcement communities must work in coordination to achieve the most effective response (Fig. 1). More creative strategies for both vertical and horizontal communication among the involved biosecurity and law enforcement agencies are needed (Ochoa-Corona 2011).

What Is Forensics and How Does It Differ from Traditional Plant Disease Diagnosis and Traceback?

Traditional plant disease diagnosticians use multi-faceted approaches to detect and identify diseases and to identify the causal pathogens. Their primary stakeholders include producers and farmers, extension educators, crop consultants, and regulatory officials, and the immediate and primary goal of traditional diagnostics is to identify the pathogen to the taxonomic level required to recommend effective means of disease management, thereby limiting the impacts of the outbreak. Longer term objectives often include understanding the epidemiology, progression and spread of the plant disease and prevention of future outbreaks.

Strictly speaking, forensics is the application of scientific methods and strategies to solve a crime, with the primary goal of connecting the crime to a perpetrator (or perpetrators) for the purpose of criminal attribution (Budowle 2003; Budowle et al. 2003, 2005a, b, c, d). The major stakeholders of forensic science include members of the law enforcement, security, investigative and regulatory communities. Because forensic evidence is often presented in a court of law, where it is subjected to rigorous and critical challenge, forensic investigatory methods such as sampling strategies, diagnostic and pathogen identification and discrimination assays, and other

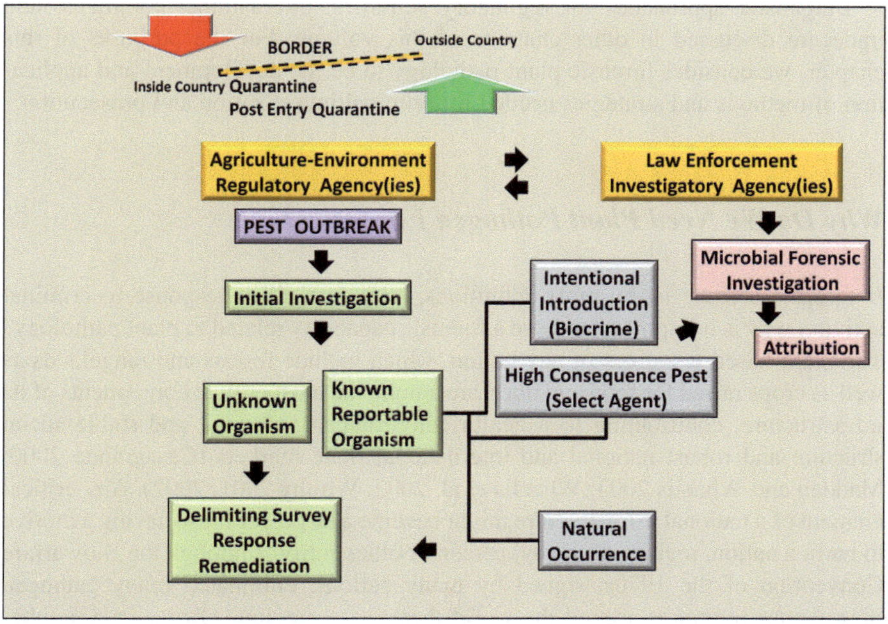

Fig. 1 Interactions between multidisciplinary teams from the diagnostic, regulatory (agriculture-environment and regulatory, or AERA), and law enforcement (LE) communities, working coordinately to achieve the most effective response. A biocrime can occur either at the border or within a country, and biosecurity violations are often committed unintentionally. They occur frequently at the border, where they are handled by the relevant local AERA, such as the Department of Homeland Security's Customs and Border Protection agency in the United States, which works with the support of the USDA's Animal and Plant Health Inspection Service, Plant Protection and Quarantine (USDA PPQ). Within a country's borders, detection of a pathogen that is exotic, or that in some other way prompts suspicion of criminal activity, frequently prompts an initial investigation led by an AERA (in the United States, the USDA PPQ) that may lead to further decision making. If the identity of the organism is unknown the pest is identified by an AERA. Any of a number of scenarios may occur after the pest is fully identified. If it is a reportable organism (including high consequence pests, known in the United States as "select agents") and/or a criminal intervention is suspected, then the appropriate LE investigatory agency may be contacted by the AERA to determine whether a crime has occurred. If there is evidence of intentional manipulation or introduction of an unwanted pest, a microbial forensic investigation will be initiated by LE and the results used to support attribution. However, if the case is ruled a natural occurrence it will return to the AERA, which will design and execute a delimiting survey to determine the extent of the incursion and the response required to eradicate or contain the new pest and remediate damage, if feasible

investigative approaches must be highly validated and stringently controlled by working under third party accredited frameworks (i.e. ISO-accreditations), and their parameters and limitations known. Further, some types of forensic evidence not usually considered in traditional plant disease diagnostics, such as human fingerprinting and other physical evidence, victim circumstances, possible suspect motives, and other elements, also are important.

Diagnostic approaches for regulatory activities, international commerce and trade are discussed in other chapters of this volume. For the purposes of this chapter, we consider forensic plant pathology to be the development and application of methods and strategies needed for criminal investigation and prosecution.

Why Do We Need Plant Pathogen Forensics?

If forensic science is, by most definitions, an investigative response to criminal activity, why is it important to have a forensic capability related to plant pathology? The plant-based resources of any nation, which include forests and rangelands as well as crops raised for food and fiber, are among the most critical components of its infrastructure, contributing to a healthy environment, a sound and stable social structure and robust national and international trade markets (Casagrande 2000; Madden and Wheelis 2003; Wheelis et al. 2002; Whitby 2001, 2002). Any critical element of a national infrastructure might become a target for those having a motive to harm a nation, region, company, person or other entity. Although the Biowarfare Convention of the 1970s, signed by many nations, eliminated many pathogen bioweapons programs around the world there were previous efforts, in a number of countries, to develop pathogen systems effective for this purpose (Casagrande 2000; Center for Infectious Disease Research and Policy 2003; Madden and Wheelis 2003; Wheelis et al. 2002; Whitby 2002).

Crimes are not always committed intentionally. Negative consequences (disease, un-warranted expense, quarantine, loss of harvest or of markets, etc.) can result from unintentional actions, or lack of appropriate actions. For example, if a grower purchases seed that is certified as disease-free, but later, after losing the crop to disease, learns that the order was filled inadvertently with pathogen-contaminated seed, there may be cause for a lawsuit based on criminal negligence. A rigorous forensic investigation, demonstrating that the pathogen strain obtained from plant samples collected in the farmer's field is a clear taxonomic "match" to that found in seed remaining at the provider company, would be needed. Other crimes may include multiple elements; for example, smuggling of exotic plant material such as seeds, fruits, or propagative plant parts is a frequent biocrime at airports and ports of entry, but if the smuggled material is contaminated (often cryptically) with pathogens or other exotic microbes there may be additional criminal charges for which forensic investigation is needed.

Emergence of Forensic Plant Pathology in the United States

Although microbial traceback technologies have been widely used for many years by human, animal and plant disease diagnosticians, epidemiologists, and crime investigators, interest in and development of microbial forensic capability in the

United States burgeoned in the years following 2001, when a series of letters containing viable anthrax spores were sent to several private individuals as well as to members of the United States Senate (U.S. Department of Justice 2010). Three people died from inhalational anthrax and several others became ill, but survived. This was not the first time that microbes had been used to inflict disease on others, but the 2001 U.S. anthrax incident highlighted limitations of the forensic microbial methods available at the time and spurred strong efforts to develop new and more effective methods in forensic microbiology. Appropriately, forensic methods related to human pathogens had highest priority for development, but it was recognized that forensic capacity building was needed also for other critical national infrastructures, such as agriculture, and efforts broadened to include the development of capacity related to animal and plant pathogens as well (Fletcher et al. 2006, 2010).

In 2002 a team of plant pathologists having a selected range of expertise in fields related to forensic science (disease diagnostics and epidemiology, microbial biology, genetics and evolution, statistics, and other areas), working together with forensic scientists, assessed existing plant pathology capabilities that could be applied in forensic settings. A resulting review (Fletcher et al. 2006) presents these capabilities, identifies gaps in knowledge and technology, and prioritizes needs for research, communication and extension, establishing a framework for future work. The needs fall into two areas: (1) those already being addressed, or planned, by plant pathologists in existing plant pathology research programs, and (2) those so specialized to forensics, or so rigorous for defense and justification in a courtroom, that they will require dedicated research and funding.

In 2007 the National Institute for Microbial Forensics & Food and Agricultural Biosecurity (NIMFFAB), part of the Division of Agricultural Sciences & Natural Resources at Oklahoma State University, was created to conduct research, address policy issues, provide education and training programs, and contribute to Extension programs for plant pathogen forensics and agricultural biosecurity. The institute's mission is to support national and regional biosecurity, law enforcement, and regulatory communities, and the U.S. agricultural enterprise, by identifying, prioritizing and addressing issues of forensics, agricultural biosecurity and food safety and by providing educational and training opportunities for students and practitioners.

Special Issues for Forensic Diagnostics of Plant Diseases

Has a Crime Occurred?

A prerequisite to any forensic investigation is an informed judgment that a crime has been committed. We want to know if the presence of a plant pathogen or the occurrence of a plant disease may have resulted from criminal activity (Fletcher et al. 2006). Since agricultural producers and consultants, environmental specialists and plant disease diagnosticians are generally unused to considering the possibility of intentional intervention (a term we have called "suspicion inertia"), it is important to consider what features of a plant disease event might prompt a contact that would lead to investigation. The presence of plant pathogens in imported or hand-carried plant material at international airports or ports of entry is usually inadvertent, but if detected may be investigated as a case of criminal negligence on the part of a passenger(s) or as an intentional criminal introduction. In addition to traditional plant disease features such as disease distribution within the field or region, symptom type and severity, time of year, previous cropping history, previous disease history, and presence or absence of vector insects, investigators will also consider the presence of physical evidence of human intervention (sprayer equipment, footprints, etc), possible motives of acquaintances or family members, and other forensic evidence. Recently, Rogers (2011) organized these factors into an illustrated and annotated decision tool, called the for use by investigators encountering a potential crime scene.

Sampling for Plant Pathogen Forensics

If preliminary investigation of a plant disease event suggests a realistic possibility that it could have been incited intentionally, a forensic investigation will be initiated. At the site, field data such as symptoms and epidemiological information are assessed to determine appropriate sampling and analysis techniques and locations.

Statistical Perspectives. Many plant pathologists are experts at field experiment design and sampling. However, a forensic investigation involving a crop differs in important ways from a traditional crop experiment having a specified experiment design, and the differences manifest themselves in data sampling and statistical analysis. These two situations have different hypotheses and objectives, and likely require different approaches to both data collection and the methodology used to arrive at conclusions. Because many scientists are trained in traditional statistical methodology, we will contrast forensic situations with what is commonly accepted in non-forensic science.

Formulation of a hypothesis, a critical element of the scientific method, is central to most scientific endeavors. For example, in a plant pathology experiment we could test the hypothesis that the development of symptoms of *Wheat streak mosaic* in wheat is affected by the plant's location within the field (north v. south, next to water versus not, etc.). Many assumptions and conditions made in common statistical approaches are common to both traditional experiments and forensics investigation. For example, if a variable, such as crop yield, conforms to the normal assumption in a designed experiment, then this variable would also maintain that assumption in the forensic case. However, there is another important condition.

In a traditional experimental situation, samples are usually taken, in one way or another, randomly, ensuring that introduction of biases, intentional or not, is minimized by the sampling protocol. Treatments are generally assigned randomly to experimental units, and samples taken within an experimental unit (i.e., plants collected within a treatment plot) are selected at random from all possible sampling units. When randomness is not utilized, biases are more likely. For example, if we allow a physician to select which of his patients will receive an experimental treatment and which will receive a placebo, he, being primarily concerned with their well-being, would likely be selective. If relatively healthier patients receive placebos, the potential for their improvement will be much less than for the sicker patients receiving treatments. In this situation, the bias is increasing the measured treatment effect. Conversely, in the drastic case of a severe disease in which the response variable is survival time, the sicker patients will have a shorter expected survival time, even with the experimental drug, and the bias would cause an underestimation of the treatment effect.

It could be argued that we should try to avoid bias and take samples randomly in both experiments and forensic investigations. However, forensic considerations may override this requirement, and the nature of the hypotheses that we test in this scenario may make randomness difficult to attain. In fact, the frame of reference to which we draw inference is inherently different from an experimental situation. If it is our goal to determine whether an infestation in a single crop field was intentional, we are not drawing inference to a larger population, but to a much-reduced subset (the single field). This eliminates the need for the random sample and frees us to sample in a more systematic manner.

Randomness is a requisite of the t-test, regression and analysis of variance. How does the lack of randomization affect the statistical approach? Many outstanding data analysts simply assume lack of randomness is not a problem and move forward anyway. However, a more solid approach would be to use **non-parametric** (having no reliance on assumptions) tests. In particular, resampling techniques such as bootstrapping, jackknifing, and permutation tests are particularly useful when randomness is not assumed (Manly 2007). Consider a field that is suspected to have been purposefully infested with a plant pathogen. Suppose also that there is another field that we are confident was infested naturally, and we want to know if the average in-plant pathogen titer in the suspected field is the same as that in the reference field. If we have ten samples from each field, and from each sample set we estimate the pathogen titer, ordinary normal-based statistics could be applied. A t-test comparing the two sample means could be calculated, implying that these sample means are estimating population means, and that these population means are the focus of our study and the subject of our null hypothesis. However, if we are concerned that these are not random samples, per se, because these plots were not sampled randomly from a larger population, we could consider an alternative.

In the previous example, the hypothesis is nonparametric because we are testing the statistical distribution of the observed values rather than means; the hypothesis is that the distribution of the suspected samples (i.e., the range and relative spread of the data, the value of the median, etc.) is the same as the distribution of the reference samples. The method takes all possible permutations of two samples, each with ten observations, from the 20 samples originally taken (184,756 possible permutations). Computers derive a statistic for each. For this scenario, it is acceptable to use a t-test, though other values could be used. A distribution of test statistic values is created from the collection of permutations. Of interest is where, in this distribution, our observed permutation falls. In a typical outcome it will fall in the middle of the distribution, and we can conclude that the two distributions are equivalent. If there is evidence that the distributions are different, then the t-test statistic would fall in the tail of the distribution of all permutations. This method allows us to address questions regarding samples without the making the typical assumption of a random sample.

Sample Collection. Since the quality of a forensic investigation is dependent on the quality of the evidence, sample collection at the incident location is a critical step that requires advance planning and attention to detail. Low pathogen titers in plant hosts or other matrices (irrigation water, soil, etc) may be caused by an uneven distribution of microbes, seasonal/climatic factors or developmental stages of the host plant (Lebas and Ochoa-Corona 2007; Ochoa-Corona 2011). In addition to collecting plant material, many sampling strategies and tools, including swipes and Q-tips for surface swabbing, special collectors for liquids, and others, have been developed and validated for a variety of settings and sample types, but validation for the site under investigation may be important. Lateral flow collection devices and immunoassays, now becoming more widely available, offer simple methods for testing and capturing microorganisms (Wong and Tse 2009). For example, a lateral flow collection tool, designated an "elution independent collection device" (EICD), has been adapted for convenient sampling of plant pathogenic viruses, bacteria and fungi (Caasi et al. 2013).

Chain of Custody. The requirement for rigorous sample management in a forensics investigation, which may begin in a crop field, a barn, a garage or a port of entry, extends well beyond collection to include every step of sample handling and labeling, storage before and after transport, and use in laboratory assays. It is critical that the whereabouts and custodian(s) of the sample are known and documented at all times, a process, known as maintaining a **chain of custody**, generally carried out by law enforcement or federal agency personnel. Custody records verify that a sample was collected and stored under suitable conditions and that any transfer from one handler to another was authorized and appropriate, establishing that the sample's integrity and purity has not been compromised. Samples shipped by mailing services are handled by trained employees in a validated manner, and shipping records document each step of the trip.

Plant Disease Diagnostics for Microbial Forensics Applications

Features of Forensic Assays for Pathogen Detection, Identification and Discrimination

Because microbial forensics is applied, in case investigations, to generate evidence for criminal attribution and/or tracing the source of a microbe to its point of origin it is conducted within a rigid legal framework and demands rigorous (accredited, ISO 17025) and unbiased performance (Budowle et al. 2005b). Detection/diagnostic techniques used for investigative purposes must be robust enough to pass strict scrutiny in a court of law. Scientists involved in plant pathogen forensics and other plant biosecurity endeavors share many common goals with regulatory, law enforcement and government agencies. Many methods and techniques are similar to those applied in traditional plant pathology, but forensics applications demand exhaustive prior characterization.

Features of robust forensic microbial detection assays include (1) a minimum acceptable level of **sensitivity** (the ability to detect very small amounts (traces) or degraded target agents or molecules); (2) optimal **specificity** (the assay's reactivity with only the target pathogen(s)) and understanding of the specificity boundaries; (3) **reproducibility** (providing the same results when performed by different operators, or over time); and **validity** (achieved by testing the assay against a substantive list of microbes that are either taxonomically or ecologically similar to the target but with which the assay should yield negative results (**exclusivity panels**) as well as a comprehensive group of microbes within the same taxon to which the test should yield positive results (**inclusivity panels**)). Examples of microbial forensic assay development in which such validation steps were fulfilled include variations on the use of real time PCR for plant pathogenic fungi, oomycetes, bacteria, and viruses and even insect pests of plants (Arif et al. 2012, 2013; James et al. 2013; Ouyang et al. 2013).

Assay Features

Pathogen Detection Assays. Pathogen detection methods used in microbial forensics include an array of serological and molecular detection assays, mass spectrometry, nucleic acid sequencing and bioinformatics (Lebas and Ochoa-Corona 2007; Ochoa-Corona 2011; Stobbe et al. 2013). Selection of the most appropriate method (s) depends on the type of pathogen, the tools available, and the scope of the screening. The scope and end user of the screening method may also bring need for rapid processing.

In general, the specificity and high throughput of serological methods such as ELISA can be exploited to specifically target microorganisms during microbial forensics or biosecurity investigations. However, either end-point or real time polymerase chain reaction (PCR) assays have largely replaced serology in forensic applications due to their specificity and sensitivity, which facilitate the detection of minimal amounts of target DNA or RNA and support the validation of results from other, previously applied techniques. However PCR does not offer the high throughput that can be achieved with ELISA and the overall cost is higher (Lebas and Ochoa-Corona 2007; Ochoa-Corona 2011).

Detection tool selection also depends on whether the pathogen is known, suspected or completely unknown. A "pre-determined" test targets a specific known pathogen, uses a specific component (antibody or DNA oligo) and is commonly applied in one pathogen in one assay at a time (Lebas et al. 2006, 2009; Ochoa et al. 2010). Differently, a "non-predetermined" test is used when the investigator does not know the identity of the pathogen under investigation. It addresses the problem by involving more general or multiplex approaches as necessary. Multiplex PCR and serological assays have been developed for microbial forensics and biosecurity applications (Boureau et al. 2013; Charlermroj et al. 2013).

Recently, interest has turned to metagenomics strategies that offer the possibilities of identifying any (and multiple) organism(s) in a complex sample, such as a plant or soil. For example, plant pathogens, including fungi, bacteria and viruses of national importance, were detected using a massively parallel sequencing (MPS) approach that combines pyrosequencing and 'reverse BLAST' functions to generate a high volume of overlapping short sequence reads that contain sequences for all DNA sources (both host and pathogen) in a sample (Stobbe et al. 2013). This strategy streamlines metagenomic capacity for detection, identification and forensics by eliminating contig assembly and searching MPS databases using only the key diagnostic sequences of interest (Stobbe et al. 2013). Further, MPS offers the possibility of detecting common landmarks of genetically modified organisms.

Pathogen Discrimination Assays. Because microbial forensics is often a question of fine-level "matching" of microbes found at a crime scene with those associated with a suspect, microbes may be subjected to molecular fingerprinting techniques such as restriction fragment level polymorphisms (RFLP), multi-locus variable repeat assays (MLVA), single nucleotide polymorphism (SNP) assays, single sequence repeat or inter-single sequence repeat (SSR and ISSR) assays that will discriminate among strains or isolates of a pathogen. The use of SNP (single-nucleotide polymorphisms) variation occurring when a single nucleotide in the genomic region of interest differs between strains of a biological species has been used with several methodologies such as DNA sequencing, capillary electrophoresis, single-strand conformation polymorphisms (SSCP), restriction fragment length electrophoresis (RFLP), mass spectrometry, high resolution melting PCR (PCR-HRM) and others (Budowle et al. 2005c; Hopkins et al. 2007; Lindstedt 2005; van Belkum et al. 2007; Winder et al. 2011; Zhao et al. 2012).

In all cases, methods used in microbial forensics must be reliable and accurate, features achieved by systematic validation and rigorous standardization based on repeatable assay design, high confidence levels and reliable reference controls.

Evidence Interpretation and Criminal Attribution

Forensic evidence is judged in the courtroom, where attribution is based on a 'preponderance' of all types of evidence, and the judgment is conferred by a jury panel (Fletcher et al. 2010). The work of a forensic scientist follows a path very different from that of a traditional plant disease diagnostician. In the investigation of a crime that may involve plant pathogens, forensic plant pathologists must gather, safeguard, analyze and interpret a comprehensive package of information to be used by prosecutors or defendants as evidence in a court of law.

Even if diagnostic assays provide pathogen identification data having acceptable confidence levels and the same pathogen is found to be associated in some way with both the crime scene and a suspect, this evidence alone may not provide conclusive proof that a suspect was indeed the perpetrator of a crime. Interpretation of field and laboratory tests must be done in consideration of other evidence such as the chain of sample custody, the history of the disease site and crop, possible motives and access of individuals other than the suspect (including those impacted by the disease outbreak) and other relevant factors. Furthermore, even a comprehensive package of evidence and court testimony is unlikely to result in 100 % confidence in a verdict.

Capability in forensic plant pathology continues to grow, along with awareness of the importance of a comprehensive suite of technologies, strategies, and trained practitioners, to assure a strong national biosecurity framework.

References

Arif M, Ochoa-Corona FM, Opit GP, Li Z-H, Kucerová Z, Stejskal V, Yang Q-Q (2012) PCR and isothermal-based molecular identification of the stored-product psocid pest *Lepinotus reticulatus* (Psocoptera: Trogiidae). J Stored Prod Res 49:184–188

Arif M, Fletcher J, Marek SM, Melcher U, Ochoa-Corona FM (2013) Development of a rapid, sensitive and field deployable Razor Ex BioDetection system and qPCR assay for detection of *Phymatotrichopsis omnivora* using multiple gene targets. Appl Environ Microbiol 79:2312–20. doi:10.1128/AEM.03239-12

Boureau T, Kerkoud M, Chhel F, Hunault G, Darrasse A, Brin C, Durand K, Hajri A, Poussier S, Manceau C, Lardeux F, Saubion F, Jacques M-A (2013) A multiplex-PCR assay for identification of the quarantine plant pathogen *Xanthomonas axonopodis* pv. *phaseoli*. J Microbiol Meth 92:42–50

Budowle B (2003) Defining a new forensic discipline: microbial forensics. Profiles in DNA 6:7–10. [Online]. http://www.promega.com/profiles/601/ProfilesInDNA_601_07.pdf

Budowle B, Schutzer SE, Einseln A, Kelley LC, Walsh AC, Smith JA, Marrone BL, Robertson J, Campos J (2003) Public health. Building microbial forensics as a response to bioterrorism. Science 301:1852–1853

Budowle B, Burans J, Breeze RG, Wilson MR, Chakraborty R (2005a) Microbial forensics. In: Breeze RG, Budowle B, Schutzer SE (eds) Microbial forensics. Elsevier Academic Press, San Diego, pp 1–26

Budowle B, Johnson MD, Fraser CM, Leighton TJ, Murch RS, Chakraborty R (2005b) Genetic analysis and attribution of microbial forensics evidence. Crit Rev Microbiol 31:233–254

Budowle B, Murch RS, Chakraborty R (2005c) Microbial forensics: the next forensic challenge. Int J Leg Med 119:317–330

Budowle B, Schutzer SE, Ascher MS, Atlas RM, Burans JP, Chakraborty R, Dunn JJ, Fraser CM, Franz DR, Leighton TJ, Morse SA, Murch RS, Ravel J, Rock DL, Slezak TR, Velsko SP, Walsh AC, Walters RA (2005d) Toward a system of microbial forensics: from sample collection to interpretation of evidence. Appl Env Microbiol 71:2209–2213

Caasi DRJ, Arif M, Payton M, Melcher U, Winder L, Ochoa-Corona FM (2013) A multi-target, non-infectious and clonable artificial positive control for routine PCR-based assays. J Microb Meth 95:229–234

Casagrande R (2000) Biological terrorism targeted at agriculture: the threat to U.S. national security. Nonprolif Rev./Fall-Winter: p 92–105. [Online]. http://cns.miis.edu/pubs/npr/vol07/73/73casa.pdf

Center for Infectious Disease Research and Policy (2003) Overview of agricultural biosecurity. University of Minnesota. [Online]. http://www.cidrap.umn.edu/cidrap/content/biosecurity/agbiosec/biofacts/agbiooview.htm. Accessed 30 May 2003

Charlermroj R, Himananto O, Seepiban C, Kumpoosiri M, Warin N, Oplatowska M, Gajanandana O, Grant IR, Karoonuthaisiri N, Elliott CT (2013) Multiplex detection of plant pathogens using a microsphere immunoassay technology. PLoS ONE 8(4):e62344. doi:10.1371/journal.pone.0062344

Fletcher J, Bender CL, Budowle B, Cobb WT, Gold SE, Ishimaru CA, Luster DG, Melcher UK, Murch RL, Scherm H, Seem RC, Sherwood JL, Sobral B, Tolin SA (2006) Plant pathogen forensics: capabilities, needs and recommendations. Microbiol Mol Biol Rev 70:450–471. doi:10.1128/MMBR.00022-05

Fletcher J, Barnaby NG, Burans JP, Melcher U, Nutter FW Jr, Thomas C, Ochoa Corona FM (2010) Forensic plant pathology. In: Budowle B, Schutzer SE, Breeze RG, Keim PS, Morse SA (eds) Microbial forensics, 2nd edn. Academic Press-Elsevier, Burlington, VT, pp 89–105

Hopkins KL, Maguire C, Best E, Liebana E, Threlfall EJ (2007) Stability of multiple-locus variable number tandem repeats in *Salmonella enterica* serovar Typhimurium. J Clin Microbiol 45:3058–3061

James M, Blagden T, Moncrief I, Burans J, Schneider K, Fletcher J (2013) Validation of real-time PCR assays for bioforensic detection of model plant pathogens. Biosecur Bioterr (Accepted; in press)

Lebas BSM, Ochoa-Corona FM, Elliott DR, Double B, Smales T, Wilson JA (2006) Control and monitoring: quarantine situation of *Plum pox virus* in New Zealand. OEPP/EPPO Bull 36:296–301

Lebas BSM, Ochoa-Corona FM (2007) Impatiens necrotic spot virus. In: Rao GP, Bragard C, Lebas BSM (eds) Characterization, diagnosis & management of plant viruses, vol 4, Grain crops & ornamentals. Studium Press LLC, Houston, pp 221–243

Lebas BSM, Ochoa-Corona FM, Elliott DR, Tang J, Blouin AG, Timudo OE, Ganev S, Alexander BJR (2009) Investigation of an outbreak of *Soil-borne wheat mosaic virus* in New Zealand. Austral Plant Pathol 38:85–89

Lindstedt B (2005) Multiple-locus variable number tandem repeats analysis for genetic fingerprinting of pathogenic bacteria. Electrophoresis 26:2567–2582

Madden L, Wheelis M (2003) The threat of plant pathogens as weapons against U.S. crops. Annu Rev Phytopathol 41:155–176

Manly BFJ (2007) Randomization, bootstrap and Monte Carlo methods in biology. Chapman & Hall, Boca Raton

Ochoa-Corona FM (2011) Biosecurity, microbial forensics and plant pathology: education challenges, overlapping disciplines and research needs. Austral Plant Pathol 40:335–338

Ochoa-Corona FM, Tang JZ, Lebas BSM, Rubio L, Gera A, Alexander BJR (2010) Diagnosis of *Broad bean wilt virus 1* and *Verbena latent virus in Tropaeolum majus* in New Zealand. 2010. Austral Plant Pathol 39:120–124

Ouyang P, Arif M, Fletcher J, Melcher U, Ochoa-Corona FM (2013) Enhanced reliability and accuracy for field deployable bioforensic detection and discrimination of *Xylella fastidiosa* subsp. *pauca*, causal agent of citrus variegated chlorosis using Razor Ex technology and TaqMan quantitative PCR. PLoS One (Accepted; in press)

Rogers SM (2011) A decision tool and molecular typing technique for plant pathogen forensic application, using *Wheat streak mosaic virus* as a model pathogen. PhD thesis, Oklahoma State University

Stobbe T, Daniels J, Espindola A, Melcher U, Ochoa Corona F, Garzon C, Verma R, Fletcher J, Schneider W (2013) Electronic diagnostic nucleic acid analysis (EDNA): a theoretical approach for improved handling of massively parallel sequencing data for diagnostics. J Microbiol Methods http://dx.doi.org/10.1016/j.mimet.2013.07.002

United Nations' Food and Agriculture Organization (2002) Biosecurity in food and agriculture: scope and relevance. Report of the Expert Consultation on biosecurity in food and agriculture (TC/BRM 03/2), Rome, Italy

United States Department of Justice (2010) Amerithrax investigative summary. Department of Justice Website, 96 pp

van Belkum A, Tassios PT, Dijkshoorn L, Haeggman S, Cookson B, Fry NK, Fussing V, Green J, Feil E, Gerner-Smidt P (2007) Guidelines for the validation and application of typing methods for use in bacterial epidemiology. Clin Microbiol Infec 13:1–46

Wheelis M, Casagrande R, Madden LV (2002) Biological attack on agriculture: low-tech, high impact bioterrorism. Bioscience 52:569–576

Whitby SM (2001) The potential use of plant pathogens against crops. Microbes Infect 3:73–80

Whitby SM (2002) Biological warfare against crops. Palgrave, Basingstoke

Winder L, Phillips C, Richards N, Ochoa-Corona FM, Hardwick S, Vink C, Goldson S (2011) Evaluation of DNA melting analysis as a tool for species identification. Meth Ecol Evol 2:229–332

Wong R, Tse H (2009) Lateral flow immunoassays. Humana Press/Springer, New York, 223 pp

Zhao A, Poulin L, Rodriguez-R LM, Serna NF, Liu S, Wonni I, Szurek B, Verdier V, Leach JE, He Y, Feng J, Koebnik R (2012) Development of variable number of tandem repeats typing scheme for the bacterial rice pathogen Xanthomonas oryzae pv. oryzicola. Phytopathology 102:948–956

Part III
Role of Diagnostics in Plant Disease Management

Results of the EU Project QBOL, Focusing on DNA Barcoding of Quarantine Organisms, Added to an International Database (Q-Bank) on Identification of Plant Quarantine Pathogens and Relatives

Peter J.M. Bonants

Abstract The rate of introduction and establishment of damaging plant pests and diseases has increased steadily over the last century as a result of expanding globalisation of trade in plant material, climate change, EU expansion, and by a recognised decline in the resources supporting plant health activities. Furthermore there is a constant decline in the number of taxonomic specialists in the different disciplines (mycology, bacteriology, etc.), capable of identifying plant pathogens, and funds to support this kind of work are very hard to obtain. Also the number of other specialists in phytopathology and other fields, which are vital for sustaining sound public policy on phytosanitary issues, are diminishing. These problems affect all countries. In this context QBOL (www.qbol.org), an EU project on DNA barcoding, started in 2009 to generate DNA barcoding data of quarantine organisms and their taxonomically relatives to support plant health diagnostics. The data are included in a database, called Q-bank (www.Q-bank.eu), which now consists of a dynamic open-access database of quarantine plant pests and look-alikes, linked to curated and publicly accessible reference collections. It contains sequence and morphological data including photographs, nomenclatural and diagnostic data of specimens available in reference collections. Within Q-bank curators from many countries with expertise on taxonomy, phytosanitary and collection issues for the different groups have been appointed and links with other databases have been made; this in order to provide Q-bank an international role in supporting plant health agencies.

Keywords QBOL • DNA barcoding • Quarantine plant pathogens • Q-bank • Database

P.J.M. Bonants (✉)
Plant Research International, Wageningen UR, Wageningen, The Netherlands
e-mail: peter.bonants@wur.nl

M.L. Gullino and P.J.M. Bonants (eds.), *Detection and Diagnostics of Plant Pathogens*, Plant Pathology in the 21st Century 5, DOI 10.1007/978-94-017-9020-8_8,

Introduction

Development of accurate identification tools for plant pathogens and pests is vital to support European Plant Health Policies. For the EU project QBOL project (Bonants et al. 2010) Council Directive 2000/29/EC is important, listing some 300 organisms for which protective measures against introduction into and their spread within the Community needs to be taken. Those threats are now greater than ever because of the increases in the volumes, commodity types and origins of trade, the introduction of new crops, the continued expansion of the EU and the impact of climate change. Currently identifying pathogens (in particular new emerging diseases) requires a staff with specialized skills in all disciplines (mycology, bacteriology, etc.); which is only possible within big centralized laboratory facilities. Taxonomy, phytopathology and other fields which are vital for sustaining sound public policy on phytosanitary issues are threatened with extinction. Modern molecular identification/detection techniques may address the decline in skills since they often require much less specialist skills to perform, are more amenable for routine purposes and can be used for a whole range of different target organisms. Recently DNA barcoding has arisen as a robust and standardized approach to species identification. A DNA barcode is a short region of the DNA which can be used to identify a species. QBOL made DNA barcoding available for plant health diagnostics and focused on strengthening the link between traditional and molecular taxonomy as a sustainable diagnostic resource. Within QBOL, collections harboring plant pathogenic Q-organisms were made available. Informative genes from selected species on the EU Directive and EPPO lists have been DNA barcoded from vouchered specimens and the sequences, together with taxonomic features, have been included in a new internet-based database system: Q-bank: www.q-bank.eu. A validation procedure on developed protocols and the database have been undertaken across worldwide partners to ensure robustness of procedures for use in a distributed network of laboratories across Europe.

Q-bank was first developed in a Dutch FES project financed by the ministry of Economic Affairs. This project (2006–2010) aimed to strengthen the infrastructure on plant health. During this project and in some other small projects from the NPPO, universities and institutes more data was incorporated in the seven databases of Q-bank. Curators from different NPPO (national plant protection organizations), universities and institutes take care that data included meet certain standards. Important to mention is that all data are retrieved from specimens which are available in collections.

QBOL Project's Objectives

Four principal project objectives were formulated within the QBOL project and are shown below:

1. Obtain or produce relevant vouchered sequence data for individual pests or pest groups and position them in a correct taxonomic context. We will determine which and how many genes (barcodes) are informative for correct Q-species identification and what are the species limits for relevant Q-organisms and morphologically and/or taxonomically related organisms, to enable the accurate identification/diagnosis of all taxa on the EU Council Directive and EPPO A1 and A2 lists.
2. Developing generic diagnostic tools based on these barcode sequences. We will investigate bioinformatics tools to enable the correct identification of Q-organisms based on DNA barcode sequences, and develop a database that will enable the storage and searching of related diagnostic metadata, to link vouchered sequence information to published biological information.
3. Develop strategic approaches and methodologies to enable the establishment of DNA banks and access to digital voucher specimens. We will develop methods that enable the storage of DNA/RNA samples (a DNA bank) for the selected set of Q-organisms and their relatives to enable access of material to all national plant protection services for positive and negative controls
4. We will support better collaboration between EU and third-world country diagnostic laboratories and also the international 'DNA barcoding' community.

Main Results

To meet the objectives of QBOL, we performed research on the different aspects of DNA barcoding for the selected quarantine organisms, the database, DNA-bank, validation and dissemination within the different work packages. The results of all these work packages are described below:

Work Package 1 – Coordination and Management

QBOL Project activities were continuously coordinated and managed by the project team of the coordinator PRI. The webportal, which was set up for partners for internal communication between partners/Advisory Board and exchange of presentations, reports, minutes etc., was regularly updated. Seven project meetings were organized in Wageningen (May 2009), Montpellier (October 2009), York (May 2010), Bologna (October 2010), Slagelse (March/April 2011) and Gent (September 2011). The final meeting was held in Haarlem on 21 May 2012.

Work Package 2 – Barcoding Fungi

Within WP2 a short list of 19 Q-species were selected for barcoding. For those species several gene regions were screened to identify suitable barcoding loci. Protocols for efficient DNA extraction, generic amplification and sequencing of the selected loci were evaluated.

For some species it was difficult to obtain larger numbers of isolates per quarantine species. This work package succeeded to make available to the Q-bank database: 791 sequences for 193 strains from 25 quarantine species; as well as 6,107 sequences from 1,145 strains of 612 related species. Of these sequences, 360 sequences for 81 species were extracted from studies in peer-reviewed journals (mainly for the related species of the unculturable obligate biotroph genera *Melampsora*, *Puccinia* and *Thecaphora*). For each species of quarantine importance in the Q-bank database, hyperlinks to EPPO and EU Council Directive documents are provided. A field "Diagnostic locus for identification in Q-bank" is present for each quarantine species to help the end-user to determine which locus is needed for identification in the database and polyphasic identifications are possible per genus group. A molecular decision scheme (Fig. 1) showing the route to an identification starting with DNA isolation and amplification of the internal transcribed spacers (ITS) of the nrRNA operon as primary barcode is provided on the Q-bank website. A link to MycoBank, a database for the taxonomy of fungal names, is also provided for each species.

Work Package 3 – Barcoding Arthropods

Within WP3 198 species of Q-arthropods have been divided into two priority groups. These lists have been erected on the basis of the economic value of the Q-arthropods, their availability and their habitat, trying to cover both agriculture and forest pests. Many contacts with colleagues and many field trips have been made to get the required specimens.

We tested several DNA extraction methods that are commonly used with arthropods and selected three that performed better and/or are easy to use by non-trained people. Also, a non-destructive protocol was developed.

Cox-1 region and the ITS-2 region of the rDNA were selected to be barcoded and primers for those regions were developed and tested. The initial objectives were to generate from 5 to 10 barcode sequences (COI and ITS) for about 100 species of Q arthropods (priority 1), and about 50 closely related species. For priority group 1, 83 Q-species (79.8 %) were sequenced for a total of ca. 2,500 sequenced PCR amplicons with an average of 20 COI and 10 ITS sequences per species. For priority group 2, 52 Q-species (54.7 %) were sequenced for a total of ca. 1,300 sequences with an average of 16 COI and 8 ITS sequences per species.

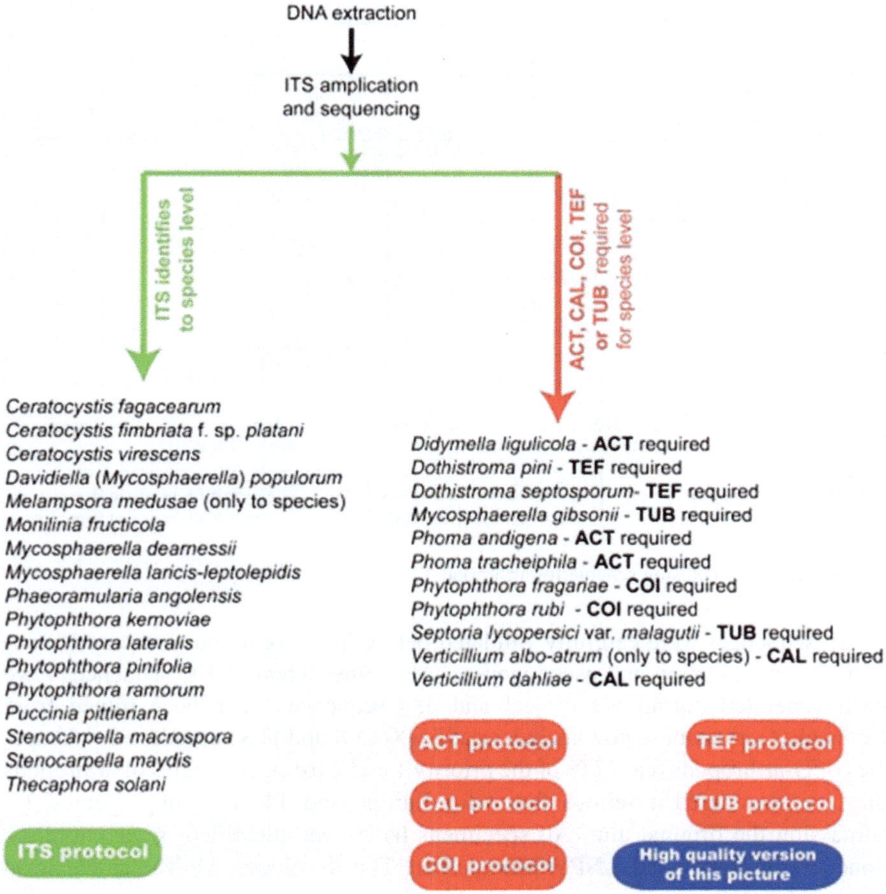

Fig. 1 Molecular decision scheme for fungi

We also included 20 species that are not yet considered quarantine species, though they represent a threat to Europe. This priority 3 list was established during the project, and we really believe that including these species make our identification tool more adapted to European needs. For these species, a total of ca. 200 sequences with an average of 7 COI and 3 ITS sequences per species were obtained. Instead of the 50 outgroup species initially proposed (i.e. species that are congeneric of or could be confused with the Q-arthropods), we sequenced 128 species, producing about 1,300 sequences with an average of 7 COI and 4 ITS sequences per species.

To improve the representativeness of our database (i.e. include Q-arthropod species included on priority 1 and 2 lists but not yet available to us or increase intraspecific variability for better identification), 334 COI sequences mostly produced by USDA and mined from GenBank have been added to our sequence library.

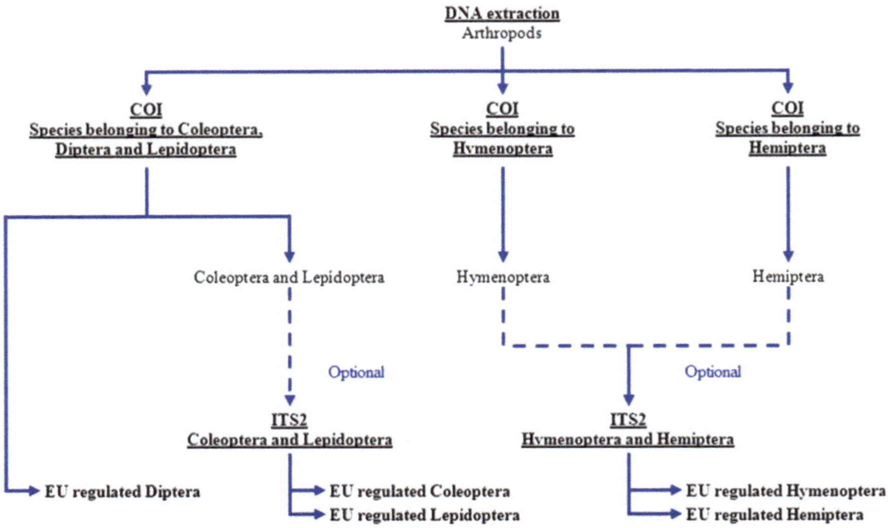

Fig. 2 Molecular decision scheme for arthropods

All sequences were carefully validated before inclusion in our database (detection of contamination and pseudogenes). Altogether, about 5,300 sequences have been generated during the project and 334 sequences have been mined from Genbank. Our database now includes ca. 5,600 COI and ITS barcodes for 153 species of Q-arthropods (ca. 77 % of the priority 1 and 2 lists), 20 species of arthropods that are considered a serious threat for Europe and 140 outgroup species, far surpassing the original aim. All specimens have been identified by taxonomists, vouchers in INRA and LNPV Montpellier. The developed Molecular Decision Scheme for Arthropods is presented in Fig. 2.

Work Package 4 – Barcoding Bacteria

Within WP4 work focussed on the Q-species within the genera *Ralstonia*, *Xylella*, *Clavibacter* and *Xanthomonas*. Most of the Q-species within *Xanthomonas* on the EU Directive and EPPO list are on the pathovar level, which makes it difficult to select the barcoding gene as pathovars do not necessarily form a single taxonomical group.

Strains were retrieved from official collections, with a substantial amount being contributed from the BCCM-LMG collection hosted at LM-UGent. Especially for the *Xanthomonas*, other official and working collections were contacted in regions of the world where the pathogens recently occurred.

DNA extraction procedures were evaluated and final protocols were written. Several genes were evaluated for their performance as a barcode region. Finally a

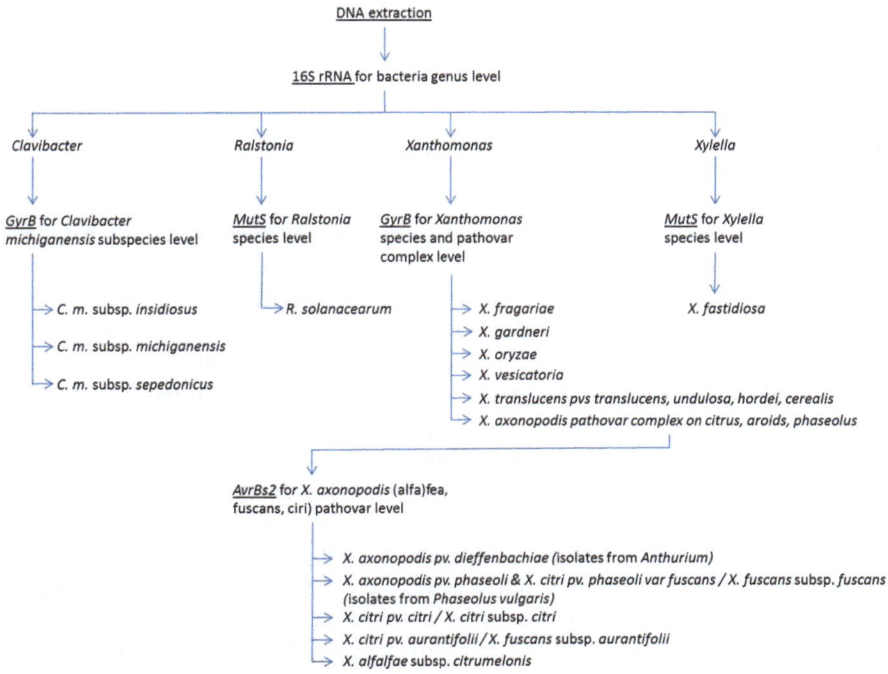

Fig. 3 Molecular decision scheme for bacteria

molecular decision scheme (Fig. 3) was published to lead the end-user through the identification process. This scheme clearly shows when to use which barcoding genes.

Extensive sequencing also confirmed the taxonomic position of most of the target Q bacteria (within *Ralstonia*, *Xylella*, *Clavibacter michiganensis* and the *Xanthomonas* species) and also supports recent proposals for the taxonomic division of the *Ralstonia solanacearum* complex. On the other hand, our research also revealed that some Q-pathogens are represented by heterogeneous strains (*e.g. X. axonopodis* pvs. *dieffenbachiae*, *phaseoli* and *allii*). Their classification under the same Q-pathovar name is questionable and needs further investigation by sequencing more genomic domains and performing host range experiments on plants. Within *Clavibacter*, the three Q- *C. michiganensis* subspecies are identified by the *gyrB*-based barcode. Also many non-pathogenic strains of the species (look-a-likes) were included in the study.

In total, the QBOL working collection increased to 1,008 strains and 3,667 sequences have been generated. Based on these sequences and on other strain characteristics (such as host, geographic origin and symptom type) a subset of reference strains has been identified for the end-users. The barcodes have been deposited in the Q-bank database and the strains are stored and available from the certified public service culture collections BCCM-LMG (BE), NCPPB (UK), and

CFBP (FR). Under the initiative of ILVO, these culture collections collaborate on a reference collection of plant health-affecting bacteria.

Work Package 5 – Barcoding Nematodes

Within WP5 a base list of 32 nematode species was created for which barcodes needed to be collected. This list contained all quarantine nematodes as well as a number of close relatives. In addition, 43 nematodes species were nominated which would be sequenced if time permitted. These additional species were composed of further relatives of Q-organisms as well as a number of other agronomically relevant nematode species. Material for most of the species on the base list has been acquired as well as material for a large number of additional species.

Five DNA isolation methods were compared, including both commercial kits as well as published methods, and the best two methods were selected for further use.

Primers were developed for the amplification of six potential barcoding regions: the small subunit (SSU) ribosomal RNA gene, the D1-D2 and D2-D3 regions of the large subunit (LSU) ribosomal RNA gene, the second intragenic spacer region (IGS2) of the ribosomal RNA cassette, a fragment of the RNA polymerase II gene and the mitochondrial cytochrome oxidase c subunit 1 (COI) and subunit 2 (COII) genes. A subset of nematode species was assigned to assess both the inter- and intra-species variation of these potential barcode regions and based on these results the SSU, LSU, COI and COII genes were chosen for sequencing in the remaining nematode species. A molecular decision scheme (Fig. 4) show the identification process. A total of 1,600 sequences for up to 58 species, distributed over the various priority groups, was promised in the project. For each priority group the required amount of sequences were generated and in some cases well surpassed. Of all generated sequences, 1,683 were of a high enough quality for inclusion in Q-Bank, originating from a total of 121 species.

Work Package 6 – Barcoding Viruses

Since viruses don't contain a generic barcode gene, we decided within WP6 to sequence the whole genome of viruses using Next Generation Sequence Technology.

The consortium produced an initial list of viral targets for which little sequence was available at the start of the project. Material was obtained for each of these species and the genome sequences were produced using 454 and Solexa technology. With the acquisition of a 454 GS-FLX by one of the partners it was possible to optimise the complete sequencing process for virus genome sequencing. Methods have now been developed within the consortium which allow the cheap combining of multiple samples prior to the sequencing processing. These methods, along with

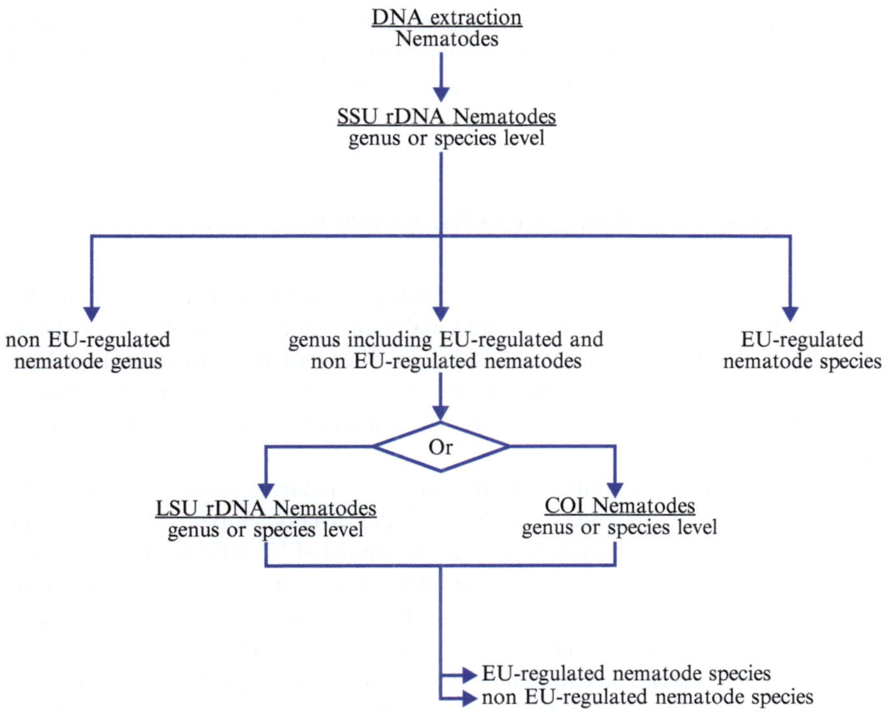

Fig. 4 Molecular decision scheme for nematodes

an optimum virus RNA specific extraction process have reduced the cost of viral genome sequencing. Genome sequence data has been produced for *Arracacha virus B, oca strain, Potato black ringspot virus, Potato virus T, Potato yellowing virus, Tomato infectious chlorosis virus, Chrysanthemum stem necrosis virus, Iris yellow spot virus, Tomato torrado virus, Tomato marchitez virus, Potato yellow vein virus* and *Tomato chocolate virus.*

A number of different RNA extraction methods have been tested and used to successfully produce virus genome sequence. It has been discovered that to maximise virus sequence recovery and thus minimise sequencing cost total RNA extraction of plants containing virus is not the best approach. Methods have now been developed within the consortium to purify virus RNA away from plant RNA. These methods have been being compared to determine the optimum method for a particular sample type. These methods include double stranded RNA isolation, small interfering RNA isolation, partial virus purification prior to RNA isolation, subtractive hybridisation and the use of capture probes. Results suggest that no one method is optimal for all samples.

A range of methods to sequence plant viruses have now been developed and validated within the consortium which allow the cheap combining of multiple samples prior to the sequencing processing. These methods along with advice on accessing this technology have now been published as part of the QBOL project.

The methods have been used to sequence in total 46 viruses and have been used in the diagnosis of a number of novel diseases including the discovery of watercress white vein virus and maize lethal necrosis in Kenyan maize.

Work Package 7 – Barcoding Phytoplasmas

Within WP7 a prioritized list of all phytopathogenic phytoplasmas relevant for the EU to be barcoded was established and maintained and expanded using existing collection with relevant phytoplasma isolates. Colleagues and collections have been contacted for specific strains throughout the project, for instance after publication of interesting new strains of phytoplasmas. To obtain new strains we mainly used the COST0807 network.

We have strains from all these Q-phytoplasmas, and have during the last months been able to obtain barcode sequences of the American type of palm lethal yellowing through a scientist in Honduras. As the list of 'Candidatus *Phytoplasma*' species is constantly expanding we are still trying to include these in the collection. We will use our international contacts to get material from these '*Candidatus*' species. However we have established a list with all 13 Q-phytoplasmas.

Phytoplasmas cannot be cultured *in vitro* and thus they need to be maintained *in planta* which requires considerable work. Partner 8 has currently 140 strains in micropropagation which covers the above-mentioned Q-phytoplasmas except palm lethal yellowing which is only available as DNA.

Several DNA extraction methods were evaluated for their effectiveness to extract phytoplasma DNA from infected host material.

Phytoplasma barcode regions 16S, tuf and SecA have been selected to be used as DNA barcodes. Tuf and SecA regions are 400–600 bp whereas the 16S region is app. 1.8 kb.

Barcode data including intra- and interspecies genetic variation on the phytoplasma barcode regions (3 regions: 16S, Tuf and SecA) were collected during the last part of the project. Until now, more than 460 barcodes have been produced. These barcodes enables good separation between phytoplasma groups and are thus ideal for identification of phytoplasmas, including quarantine organisms, as shown in the molecular decision scheme (Fig. 5).

Work Package 8 – DNA Banks

Since DNA of quarantine organisms is scarce, we investigated within WP8 different protocols to store, transport and multiply (WGA: Whole Genome Amplification) DNA samples of these organisms. Within WP8 eight different protocols for long term storage and transport of DNA/RNA samples and WGA products were

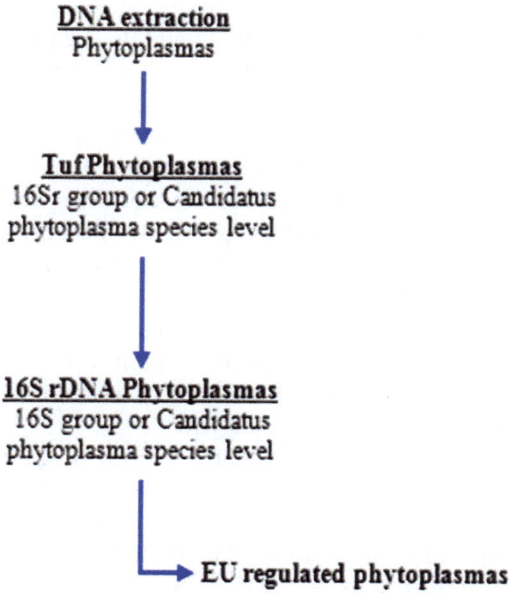

Fig. 5 Molecular decision scheme for phytoplasmas

investigated and tested (*e.g.* filter, beads, other). GenTegra was chosen as storage medium.

Four kits for whole genome amplification (WGA), a method to multiply DNA, were tested on a subset of organisms from each group (fungi, bacteria, arthropods, nematodes, viruses, phytoplasms). The quality of the individual kits was assessed using different methods: TaqMan PCR, conventional PCR, sequence analysis and gel electrophoresis. Based upon results obtained thus far a WGA kit was selected to be used for the rest of the project.

The samples for ring testing in WP10 Validation were prepared and a prototype of DNA bank was established. Several protocols have been evaluated and final protocols have been written. Using these protocols NPPO's can better handle DNA samples of rare specimen to be used as positive and negative controls in their molecular identification and detection assays.

Work Package 9 – DNA Barcode Library/Database/Informatics

The database (developed within the Dutch FES project, 2006–2010) has been further developed during the QBOL project within WP9. Together with QBOL WP leaders and associated researchers, we have created six databases: for fungi, arthropods, bacteria, nematodes, viruses and phytoplasmas. All groups have access to their databases via Citrix XenApp. The total database Q-bank (Bonants et al. 2013) is freely accessible via internet (www.q-bank.eu). Contacts with CBOL/ BOLD, EDIT WP5, GBIF, StrainInfo, GenBank and EMBL were taken at several occasions during the course of the project in order to import from and export data to the respective projects or databases. A software module to export to and import from Genbank (and therefore EMBL) has been implemented. In discussion with the curators of the database we continuously improved the Internet-based software to comply with the needs of the end-users during pairwise meetings. Additional training was provided during the meetings with WP leaders and associated researchers. Filling of the databases has been made significantly during last stages of the QBOL project but will continue even beyond the end of the project. Publication of the created databases is now complete and Internet visitors are regularly using the system. Websites are therefore not restricted to the users participating in the QBOL project anymore. Usage of the different databases are monitored by Google Analytics.

The bioinformatics and databases of Q-bank are based on the BioloMICS software (BioAware, Belgium). This tool allows specialized and scientific biological databases to be created to fit the specific needs of researchers working on any organisms (from arthropods, bacteria, fungi, nematodes, phytoplasma, plants to viruses). It is used by a broad base of users such as taxonomists, ecologists, human, or plant pathologists, molecular biologists, pharmacists, industrial researchers, etc.

Work Package 10 – Validation/Evaluation

In the first part of the QBOL project a survey was set-up to find out the wishes and expectations of possible end-users (scientists and technicians of NPPO's) in regard to the data generated by QBOL and stored into the database (Q-bank). Based on the end-users' expectations, and on the queries submitted, the usability of the QBOL database was improved.

Before the start of the test-performance study (TPS) within WP10, the developed tests were harmonised as much as possible and a draft EPPO standard "DNA barcoding as identification tool for EU regulated plant pests" has been made. A selection of specimens to be tested in TPS has been made and treated to be non-infectious and non-viable for sending them without permits. A homogeneity test has been performed with all samples before sending. Obtained sequences in the homogeneity test served as standards for comparison with the TPS outcome.

Twenty-one TPS packages (14 TPS partners, 7 training sessions) were prepared providing partners and training session organizers with an instruction booklet including de EPPO standard, all DNA purification kits, primers and samples. All results from 14 TPS partners were analysed and evaluated in terms of (1) Number of samples analysed and % of test correct used, (2) % amplicons obtained. (3) % consensus obtained, (4) % consensus sequence of correct size, (5) % primers trimmed, (6) Diagnostic sensitivity, (7) Diagnostic specificity, (8) Repeatability and (9) Robustness. Pitfalls in the use of the EPPO standard, instruction booklet and the use of Q-bank could be identified and recommendations for future work were made. A paper has been written to present the results of this TPS (van de Vossenberg et al. 2013)

Work Package 11 – Dissemination

Within the QBOL project dissemination played a major role in order to attract as much interest as possible for the outcome of this EU project. The QBOL project website (www.qbol.org) has been developed, maintained and regularly updated. The aim was to have at least 2,000 visitors per year. At the end of the project we had more than 10,000 visitors for the website. The website contains also an internal site for project participants and the Advisory Board, which contains minutes of meetings, reports, presentations, discussion forum etc., and an external site for stakeholders and end-users.

A publicity leaflet (1,000 copies) and the QBOL poster were made and distributed to stakeholders and can be obtained by the partners from the web-portal.

Participants from all WPs presented their work at (inter)national meetings and conferences via oral and poster presentations.

Seven training courses were organized: South Africa, Kenya, China, Honduras, Peru, India and the Netherlands. One hundred and thirty participants (mainly people from NPPO's: national plant protection organisations) from 26 countries attended these seven courses. In those courses the different work packages were presented. Practical work was performed on DNA extraction, barcode amplification and sequencing using the protocols developed within WP2-7 and finally searching in the database developed within WP9 (Q-bank).

A QBOL-EPPO workshop was organized in Haarlem, The Netherlands from 22 to 25 may 2012. The final QBOL workshop was held together with the EPPO meeting on Diagnostics, which takes place every 3–4 years. More than 180 people attended one or more days of the meeting. QBOL WP leaders presented the results of their work package.

During the project more regular contact with stakeholders and end-users was made. Presentations were given for heads of NPPO's, CPM (IPPC), Diagnostic panel IPPC, EPPO panel on Quality Assurance and Diagnostics, different EPPO panels, DG-Sanco and EPPO Working Party and Executive Committee.

Within the Netherlands the ministry of Economics, Agriculture and Innovation subsidised the Q-bank database for 3 years (2011–2013) to set up a long-term plan for Q-bank. A steering committee, a group of curators for the different databases, a coordinator, a program manager and the database managers work now on the quality and continuation of the database. Future incorporation of the database in EPPO activities is under discussion with EPPO and EU.

Summary

Within the QBOL project we were able to develop DNA barcodes for many plant pathogenic quarantine organisms present on the EU Directive and EPPO list and closely relatives. Protocols for DNA/RNA extraction, generic amplification of the barcoding region and sequence analysis were written and included in the molecular decision schemes, which were produced by WP2-WP7 and in which end-users can see which protocols to be used for correct identification of the quarantine organisms. Below the total number of sequences obtained is presented:

	Sequences obtained		3-10-2012
QBOL		# sequences	Remark
WP2	Fungi	6,898	8 loci
WP3	Arthropods	5,300	2 loci
WP4	Bacteria	3,667	20 loci
WP5	Nematodes	1,683	6 loci
WP6	Viruses	46	Whole genome seq
WP7	Phytoplasms	472	3 loci
	Total	18,066	

The developed database, Q-bank, is freely accessible via internet (www.q-bank.eu). Tools have been provided how to search the database and perform BLAST analysis or even multilocus identification.

Many dissemination activities have been performed (website, E-newsletters, flyers, poster, oral and poster presentations at conferences worldwide, training course in seven countries, publications in refereed journals, etc.). There was much interest from all over the world in the QBOL project and its achievements. The results were presented at the final workshop which was organized together with EPPO in Haarlem, The Netherlands (22–25 May 2012).

Q-bank has been developed in the Netherlands in a project to strengthen plant health. The project started in 2005 and finished in 2010 and has been funded by the Dutch Government. From 2009 through 2013, more than 18,000 'DNA barcodes' of relevant quarantine species and their taxonomic relatives were generated and included into Q-bank by the EU funded project, QBOL (Quarantine Barcode of Life; www.qbol.org). The Q-bank database (www.q-bank.eu) contains information on regulated plant pests and close taxonomic relatives, including invasive plant species, linked to curated and publicly accessible collections. The database contains sequence data and information on morphological features including photographs as well as nomenclatural and diagnostic information. In addition, the reference collections comprise voucher specimens, tissue and whole genome amplified DNA/RNA samples.

For many strains/specimens one or more of the following information is available:

- General information, including distribution, biological, ecological, literature references
- Collection data, including geographical data, collection type, status, specimens and/or isolates;
- Nomenclatural data, including taxonomy, systematic position;
- Morphological data, including photographs, characteristics of various taxonomical levels;
- Diagnostic data, including description of symptoms, illustrations, protocols;
- Sequence data, including sequences of barcode genes relevant to a specific group.

For part of the species (groups) the following aspects are included:

- Information on the evaluation of phytosanitary risks (Invasive Plants, partly Insects);
- Possibilities to compare sequence data and morphological features (Fungi);
- Digital image driven keys for identification (Invasive Plants);
- Information on identification and detection methods and on vectored transmission (Viruses).

Plans for continuation of the Q-bank database have been made and are now being discussed. The features and content of Q-bank makes this database a valuable tool for the correct identification of plant pathogenic quarantine and closely related organisms for now and in the future. Q-bank at the moment is still growing with relevant information on plant pathogenic organisms present in public available collections. People are invited to collaborate with the curators to increase and improve the content of the Q-bank database even further. Those interested in collaborating with Q-bank are invited to contact the coordinator. More information is available at www.q-bank.eu.

Acknowledgements The author want to thank all participants of the QBOL project and curators of the different databases for their valuable contribution to Q-bank. The financial contribution for Q-bank by the Dutch ministry of Economic Affairs and the financial support by the European Union for funding the QBOL project (project number 226482), is highly appreciated.

References

Bonants P, Groenewald E, Rasplus JY, Maes M, de Vos P, Frey J, Boonham N, Nicolaisen M, Bertacini A, Robert V, Barker I, Kox L, Ravnikar M, Tomankova K, Caffier D, Li M, Armstrong K, Freitas-Astua J, Stefani E, Cubero J, Mostert L (2010) QBOL: a new EU project focusing on DNA barcoding of quarantine organisms. Bull OEPP/EPPO Bull 40:1–5

Bonants P, Edema M, Robert V (2013) Q-bank, a database with information for identification of plant quarantine plant pest and diseases. Bull OEPP/EPPO Bull 43(2):211–215

van de Vossenberg BTLH, Westenberg M, Bonants PJM (2013) DNA barcoding as an identification tool for selected EU-regulated plant pests: an international collaborative test performance study among 14 laboratories. Bull OEPP/EPPO Bull 43(2):216–228

On-Site Testing: Moving Decision Making from the Lab to the Field

Neil Boonham

Abstract On-site testing is a term that is often used to describe two distinct activities, firstly detection is the initial locating of the pest or pathogen infected sample which in most instances is performed visually. The second activity is identification, usually this is achieved by sending suspected samples to a laboratory. In recent years there has been significant research activity in each of the areas to provide technological solutions to enable more rapid decision making. Of course it is not necessarily just inspection services who benefit from these techniques, they can be deployed throughout the farm to fork, agri-production chain by seed producers, growers, processors, pack-houses etc. to limit losses caused by pathogens and pests. How best to deploy detection methods however may provide a potential conundrum for policy makers and other stakeholders. Deploying simplified detection and identification methods remotely helps to speed up inspection and facilitates trade. However, without care this approach may risk a blinkered, targeted inspection approach and a 'winding down' of laboratory expertise which is needed during outbreaks of new pests.

Keywords Field-testing • Detection • LAMP • Acoustics • Volatiles • Remote imaging • Inspection

Introduction

Performing diagnostics of any kind is part of a decision making process which in the case of plant pests and diseases it is usually done to prevent or limit spread. In the case of notifiable pests this is usually to preventing incursion into a new area and the action taken is usually destruction of infested consignments. In the case of

N. Boonham (✉)
Plant Protection Programme, Food and Environment Research Agency,
Sand Hutton, York YO41 1LZ, UK
e-mail: neil.boonham@fera.gsi.gov.uk

M.L. Gullino and P.J.M. Bonants (eds.), *Detection and Diagnostics of Plant Pathogens*, Plant Pathology in the 21st Century 5, DOI 10.1007/978-94-017-9020-8_9,

non-notifiable pests, other measures are taken (e.g. spraying with plant protection products) which in some cases may be to eradicate the pest, but more usually actions are targeted to ameliorating its impact. In many situations the faster the decisions are made the more effective the action may be. Traditionally once a potential pest or disease has been located, samples are sent to a laboratory for testing, the delays caused by transit and in some cases laboratory testing procedures divorces those making decisions from the results, causing a delay in any actions subsequently taken. Furthermore, pests often go unnoticed at low levels of infestation or in the case of pathogens, pre-symptomatic infection stages. This failure to visually observe the problem can lead to it spreading unchecked until it has built up to such a level that it can be seen. These issues have led the drive to develop technological solutions that would fulfill two complementary roles. Firstly, and perhaps most simply, putting tools into the hands of those on the front line to enable rapid identification of pests would prevent delays in the process caused by reliance on a laboratory service. Secondly and perhaps a greater technological challenge is the development of detection tools that guide those on the front line to the site of the problem, be that the location of pre-symptomatic infection or low level of infestation of a pest. Linked together these tools enable a more efficient process by which greater efforts are focused on the most risky samples, enabling faster deployment of control measures.

In-Field Identification

Putting tools into the hands of those in the field is not a new approach, methods based on latex agglutination (Fig. 1a) have been performed for plant diseases since the early 1980s (e.g. Potato virus test kits were marketed by Ani Biotech, Finland). Since then more refined methods have been developed exploiting pathogen and in some cases pest specific antibodies to enable rapid identification. Early test kits based on latex agglutination on glass or plastic slides, whilst effective at generating a result did not give many concessions to the practicalities of someone working in the field. The methods required a large number of temperature labile reagents, had multiple steps in which reagents were added sequentially and the interpretation of the result was often subjective, requiring a fair amount of training and experience to reproduce effectively. Further developments sought to improve the usability of the methods, applying engineering solutions to the problem of sequential reagent addition. Some of these second generation kits (e.g. Alert kits by Neogen) also incorporated chemical substrates, effectively recreating laboratory ELISA methodology, yet performed rapidly on a solid support. This provided advantages in terms of both usability and also interpretation of results which were no longer subjective and easily interpreted by non-specialists in a field situation. The most significant innovation came in the late 1990s with the application of homogeneous test kit formats developed in the bio-medical arena and exploited most notably for hormone detection in pregnancy testing applications. The Lateral Flow Device

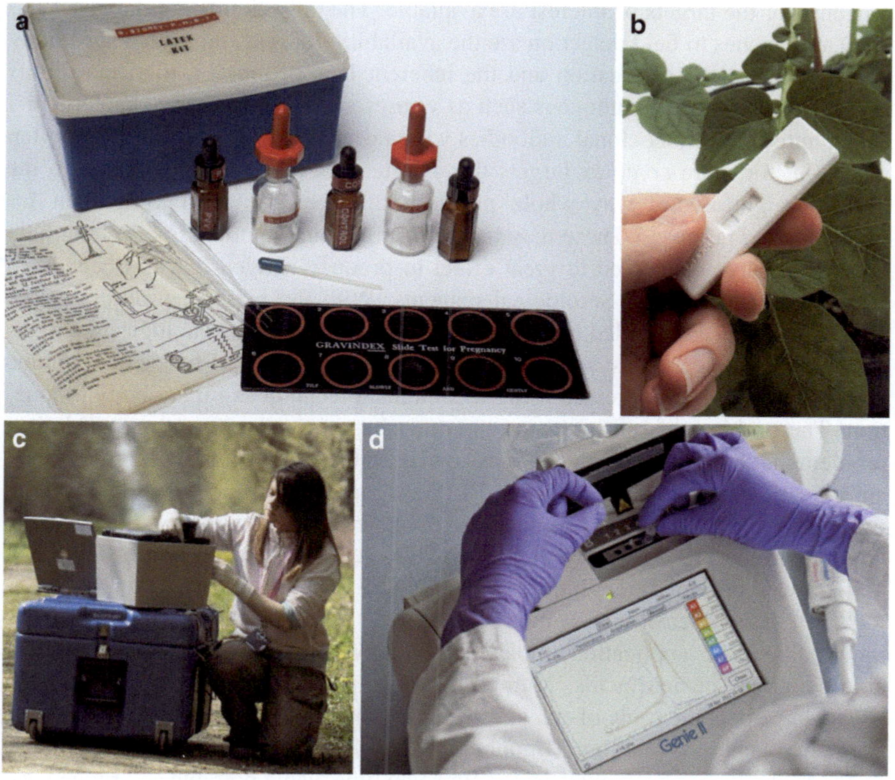

Fig. 1 Field test kits based on (**a**) Latex agglutination (**b**) Lateral flow devices (Forsite Diagnostics ltd.) (**c**) Portable real-time PCR on the smart-cycler platform (Cepheid) and (**d**) LAMP on the Genie II platform (Optigene)

(LFD) format (Fig. 1b) was exploited initially in the phytodiagnostics arena for the detection of potato viruses for use in seed certification systems (Danks and Barker 2000) and proved to be a considerable improvement over previous formats. The underlying chemistry in an LFD is effectively the same as a latex agglutination kit, the accumulation of antibody coated latex (or colloidal gold) particles caused by the presence of the antibody target. The key difference however is that the binding occurs during the capillary flow of sample and reagents along a membrane, rather than in solution. Furthermore on an LFD, the agglutination is accumulated at a specific location by the presence of a line of target specific antibody, which immobilises the agglutinated latex whilst allowing the background reagents to be washed away by continued flow along the membrane. Taken together this provides a non-subjective and clear read out of a positive result against a low background. Not only does the LFD format result in clear result, the sequential rehydration of reagents as the sample flows along the membrane effectively removes the need for multiple steps to be performed by the user. Testing based on LFD technology remains the simplest and most rapid option for field use where specific binding reagents for the targets of interest are available. The only significant drawbacks to LFD approaches to field detection are the availability of reagents with a specificity appropriate for the application and the inherent lack of amplification that limits sensitivity. For simple pathogens such as viruses and to some degree bacteria and fungi, antisera or monoclonal antibodies with a useful level of specificity are often available, for more complex targets this is often not the case, in part due to the rather straightforward way whole purified pathogens are used as antigen for immunisation purposes. These may not be insurmountable problems, recombinant techniques have been used to great effect to produce antigen that has been used subsequently to produce serological reagents. With the massive proliferation in genome sequences afforded by next generation sequencing technologies, comparative genomic techniques will in the near future be exploited to identify unique targets that could be expressed using recombinant techniques for antibody production. The problem of sensitivity on the other hand is a significant one. This is a problem that has yet to be solved for assays based on serological reagents, since any amplification achieved (e.g. AMPAK ELISA signal amplification kits) post binding typically increases the background signals concomitantly with the those of the target (Torrance 1987). This typically provides scope for further optimisation, but may provide only marginal improvements in end point sensitivity.

When greater sensitivity or more control over specificity is required in a laboratory diagnostic application, molecular biology methods are typically used to amplify target nucleic acids. Methods based on the polymerase chain reaction (PCR) have become tools-of-trade, especially when implemented with a real-time chemistry such as TaqMan (Boonham et al. 2008). Implementation of these methods on-site has been investigated for some time, using portable and ruggedised implementations of laboratory equipment. Although a number of companies have produced fieldable real-time PCR equipment (e.g. BioSeq from Smiths Detection and T-COR4 from Tetracore), the most widely used has been the Smart Cycler II produced by Cepheid (Fig. 1c). Whilst portable real-time PCR has been evaluated extensively there are a couple of significant drawbacks to it implementation. Firstly, in common with any PCR methods, extraction of nucleic acid generally requires reasonably elaborate extraction methods to avoid co-purification of compounds which inhibit the enzymes, though this has been improved for some targets (Tomlinson et al. 2010a). Secondly, whilst rugged, portable and in some cases battery powered equipment is available it remains expensive, largely due to the need for careful temperature regulation and sensitive detectors needed to record the small changes in florescence provided by TaqMan probe cleavage. To solve both of these problems subsequent research has been focused on evaluation of isothermal amplification chemistries.

Isothermal amplification as the name suggests described methods in which the amplification reaction is incubated at a single temperature. This gives advantages in terms of simplicity over PCR, since the reactions do not need to be cycled accurately between temperatures, thus water-baths, dry-blocks or incubators can be used to incubate reactions. Potentially of greater significance however are the enzymes that are used to copy DNA in isothermal reactions. These tend to be highly-processive, copying very large amounts of DNA efficiently as well as being robust and able to withstand the effects of inhibitors (Tomlinson et al. 2010b) that would prevent amplification by Taq polymerase. These factors all combine to make reactions that are ideally suited to diagnostic use, being robust, suited to testing crude DNA samples extracted from field material and producing large amounts of template which gives many options for resolution of the results.

Unlike PCR there are many different types of isothermal amplification chemistries (e.g. NASBA, RPA, HDA, SIBA), based around different mechanisms and approaches to separating DNA strands to provide single stranded binding sites for the oligonucleotide primers. Although a published systematic comparison of a large number of chemistries does not exist, by weight of published papers detailing plant pathogen assays alone (reviewed in Tomlinson and Boonham 2008) the Loop mediated AMPlification (LAMP) method seems to be the most widely adopted to date. A LAMP reaction consists of three pairs of primers (internal, external and loop primers), to generate an amplification product which contains single-stranded loop regions to which primers can bind (Notomi et al. 2000). The internal primers introduce self-complementarity into the amplification product, causing loops to form onto which primers can bind, extension of the external primers causes displacement of the products primed by the internal primers. The addition of loop or stem primers (Gandelman et al. 2011) accelerates amplification by exploiting further primer binding sites within the LAMP amplification product, enabling faster amplification and greater sensitivity. Typically LAMP can be used to achieve similar levels of specificity to PCR based assays and with a sensitivity approaching that of real-time PCR (Tomlinson et al. 2007). The products of LAMP reactions consist of alternately oriented repeats of the target sequence, which increase in size as successive rounds of copying and displacement produce a concatenated amplification product. The amount of DNA amplified within a LAMP reaction can be exploited to enable simple product detection methods but can also create a significant post-amplification contamination risk if enough care is not taken to control it.

Following incubation positive amplification can for some assays be observed directly due to the precipitation of magnesium pyrophosphate which causes a measurable increase in turbidity at the end of the reaction, although in practice the precipitate needs to be centrifuged to be routinely observed without the use of an instrument. Addition of calcein and $MnCl_2$, causes a colour change from orange to green upon amplification, whilst addition of hydroxy naphthol blue (HNB), results in a colour change from violet to blue. However all of these methods cause relatively subtle responses and are at best subjective between different users and limit the end point sensitivity of the reactions. Less subjective methods include the addition of intercalating dyes such as SYBR Green and PicoGreen at sufficiently high concentrations to produce a significant visible colour change. Incorporating ligands into amplification products such that the products can be detected in an LFD immunoassay at the end of the reaction (Tomlinson et al. 2010c; James et al. 2010) results in a test that is considerably easier to interpret. However, these methods all require the user to open tubes which not only adds steps to the test, but also pose a significant contamination risk which could result in false positive reactions.

The simplest approach to performing LAMP reactions is to use a dedicated electronic, closed tube amplification platform. An interesting parallel with laboratory PCR testing and the early developments of LAMP deployment are evident. The use of closed tube fluorescent real-time instruments resulted in the wider adoption of PCR technology into the routine diagnostic laboratory, despite the increase in cost of the equipment needed. The instruments provided a homogeneous amplification and detection platform which effectively solved the issues that had plagued diagnostic PCR uptake, that is they gave non-subjective interpretation of results from within a closed tube which limited the problem of post-PCR contamination. For LAMP, platforms that meet the requirements of end users are becoming available, and despite the research activity focused on providing an alternative will, as with PCR, provide the most effective route to routine use. The most established of these is the Genie II platform (Optigene). The platform (Fig. 1d) is a battery powered, stand alone unit that does not require connection to a computer to enable it to function. In addition it was designed specifically for diagnostics with a non-specialist user in mind, thus the touch screen interface is simple and intuitive to use and enables rapid set up and simple interpretation of the results. Typically assays can be completed within 5–20 min depending on the type of target and the nature of the sample, and for identification work simple homogenisation into an extraction buffer is the only sample processing required.

In-field testing using LFDs and simple LAMP procedures offer people working on the front line an approach to identify pests and pathogens quickly such that decision making is not held up. The methods require knowledge of the pathogen or pest to be used effectively, and in most cases are not suitable for pre-symptomatic or latent detection where bulk samples are taken and the in-field setting limits the amount of sample pre-treatment that can be performed. However for identification of individual isolated pests and for confirmation of the causal agent of disease in symptomatic material they prove extremely effective.

Finding Pathogens and Pests

Identification of pathogens and pests can only be achieved when the location of the pest or the infected material is ascertained, this is usually done by visual observation, though there are a number of technologies that may also be useful.

Sniffing Pathogen Infection

Interactions between pathogens, pests and the plant host often results in the release of volatile organic compounds (VOC) into the air (Jansen et al. 2011). Some of the

VOCs released are from the pathogen (in the case of bacteria and fungi) or pest, whilst others are from the host, many are common to both biotic and abiotic stresses and as such are non-specific (Jansen et al. 2009), yet identification of these volatile compounds can be utilised for monitoring the health status of plants (especially within containment) and potentially guiding inspectors towards the location of infected/infested material. When it comes to pest infestation it is probable that in some situations a unique VOC signature is released, even in the absence of a unique compound. For example, it has been shown that invertebrate predators have a preference for leaves infested with their specific prey compared with non-prey species (de Boer et al. 2008) despite GC-MS analysis demonstrating the individual VOCs released were identical. As a result methods that are able to discriminate patterns of VOCs rather than specific compounds may be of more value as a detection method. Sniffer dogs can perceive signatures and patterns of VOCs as well as unique compounds and have been used as far back as the 1970s for the detection of insect pests, such as Gypsy moths (Lymantria dispar) (Wallner and Ellis 1976). More recently infestation of trees and wood packaging material by Asian and citrus long horn beetles (*Anoplophora glabripennis* and *Anoplophora chinensis* respectively) have been found using trained sniffer dogs (http://bfw.ac.at/rz/bfwcms.web?dok=9531). Other animals have been used for detecting pathogen VOCs most notably insect pests; detached antennae of the Colorado beetle has been used to detect pest and pathogen infested potato material (Weißbecker et al. 1997) and intact bees can be trained to detect VOCs for example produced by infestation with fruit flies (Chamberlain et al. 2012). The problem with the invertebrate based methods is the longevity of the innate or trained response; the detached antennae have too limited a life to be practical; although live bees are a more practical prospect, the logistics of having a hive and training material to hand daily to be deployed by an inspection service would require some careful planning.

A more practical approach would be using an instrument for the detection of VOCs, most approaches to achieving this aim have three common steps, that is collection and concentration of head-space (the air containing the VOC), identification of the individual VOCs or pattern of VOCs and finally statistical analysis usually using multivariate methods to discriminate changes from the healthy status. Gas chromatography (GC) is often used for the characterisation of VOCs, and is a sensitive and specific technique. GC instruments are typically large and sensitive thus it is more practical to collect samples and analyse them in the laboratory; though small portable GC instruments have been used for the detection of VOCs from insect infested plants (Miresmailli et al. 2010). More recently electronic noses (E-nose) have been used for the detection of VOCs, these instruments are typically smaller and more portable than GC, but with a lower sensitivity. The E-nose consists of an array of sensors, each of which reacts to different volatile compound, as such they are ideal for investigating patterns of VOC rather than identifying individual compounds. Thus the pattern of reactivity of the sensor array generates a characteristic fingerprint of the VOCs associated with disease, enabling recognition of changes from healthy to diseased state using statistical techniques. The approach has been used to detect VOC patterns associated with disease of fruit tress associated with bacterial infection (Spinelli et al. 2012). The key challenges hampering the wide use of E-nose and other instrument based VOC analysis techniques are the practicalities of trapping headspace, the relatively low levels of VOCs produced and the similarity of the profile of VOCs between diseased and healthy plants and the relatively low sensitivity of techniques such as E-nose.

Listening for Wood Boring Pests

An approach that is likely to be rather specific in scope but useful for wood boring beetle larvae is the detection of the sounds made as the larvae chew through the hard fibrous substrate. Used as early as the 1930s for the detection of beetle larvae in timber and timber structures (Schwarz et al. 1935; Colebrook 1937). Acoustic methods have more recently been used to detect the larvae of the quarantine pests *Anoplophora glabripennis* and *Rhynchophorus ferrungineus* using piezoelectric sensors either attached firmly onto the surface of the wood material or inserted inside of it (Mankin et al. 2008 and Potamitis et al. 2008 respectively) and have been implemented as low cost sensors for use by inspectors (Chesmore and Schofield 2010). In addition to acoustic detection the vibrations of insects can be detected using a laser vibrometer (Andrea et al. 2008) which may offer specific advantages over the use of other sound sensors. No contact needs to be made with the substrate allowing detection of vibrations at a distance from the instrument, this may be beneficial for recording vibrations in the canopy of a tree or for valuable specimen trees (e.g. bonsai trees) where attaching the sensors using screws or inserting recording instruments into holes drilled into the sample may not be acceptable.

Seeing Infection from a Distance

Remote imaging is a catch all term for methods of visual assessment of crops, plants or the environment from a distance. These methods can be simple for example subjective assessment of aerial photos to complex methods involving analysis of spectral signatures in satellite images. Remote imaging is not a new technique, the first records detail visual assessment of aerial photographs of cotton fields where cotton root rot caused by *Phymatotrichopsis omnivora* can be clearly seen as distinct patches of contrast between the living and dead plants (Steddom et al. 2005). The earliest use of spectral analysis used infrared film where it was noted that healthy tissue reflected more infrared light than necrotic tissues, enabling the observation of symptoms before they were visible by the naked eye (Bawden 1933). Taken further, ratios of reflectance at different wavelengths (known as Vegetative Indices) offer less variation than reflectance at a single wavelength and have been used to characterise vegetation in the environment (most commonly

Normalised Difference Vegetative Index – NDVI) as well as to discriminate infected from healthy plants. For example the differences between health and infected wheat plants inoculated with stem and stripe rust (*Puccinia graminis* and *Puccinia striiformis*) was evident in advance of visual symptoms (Sharp et al. 1985). Multispectral imaging records each pixel in an image at a limited number of broad spectral bands whilst hyperspectral imaging captures many more narrower bands, both have been used to identify infection at a crop level (e.g. Franke and Menz 2007; Apan et al. 2004). Historically the data in these studies has been collected using cameras or instruments on board airplanes, more recently satellite imagery has been exploited, though purchasing images can prove expensive and may not be available for all locations. Recent developments in low cost unmanned aerial vehicles (UAV) and light-weight instrumentation may prove to be the most useful platform for performing cost effective and custom assessments in agricultural applications (Lelong et al. 2008).

Other Surveillance Tools

There are other tools that are not strictly detection methods but are vital in surveillance campaigns to provide early warning of the presence of a pest. Probably the most well used of these is the deployment of traps for insect pests. These can be simple and relatively non-specific in nature such as yellow sticky traps used in glasshouses or highly specific, using lures to attract very specific species. The traps, lures and methods used for trapping exotic insect pests have been reviewed recently (Augustin et al. 2012a, b; Quilici et al. 2012) and so detail is not given here.

One of the problems of deploying an extensive trapping network especially in the environment (e.g. for forest pests) is a logistical one. Visiting the traps regularly to assess trap catches requires significant resource, yet for regulated pests the traps are frequently empty or contain only native species. Recent work exploiting developments in wireless communication and digital imagery has however provided a potential solution (Chinellato et al. 2013; www.spensatech.com). The traps are integrated with digital cameras and the images are accessible remotely via the internet or smartphone apps, this enables wider deployment with less resource required to access the data. Additionally more traps can be deployed and contents viewed more frequently which potentially could lead to earlier detection of insect pests than is possible with a more limited network.

Conclusions

Locating pathogens and pests ahead of identifying what they are is arguably a more challenging proposition. Typically the only tool available is visual observation looking for signs and symptoms, For a detection method to be effective, sensitivity is key to enabling the user to locate infection at low incidence rates or hidden from view. Specificity on the other hand is less important; if the infected material can be found then LFD or LAMP techniques can be used to identify the cause. Another common feature of detection techniques is that they tend to be narrow in scope and only useful for individual or closely-related pests or pathogens. This has probably hampered their development as investing in non-generic methods provides limited return on investment.

Some of the barriers to the adoption of these technologies are common to methods used in the laboratory, whilst others are specific to field techniques. One of the key issues for the adoption of new methods such as LAMP is the availability of validation data in a standardised format that can be recognised by other testing laboratories. Fortunately an international standard (PM 7/98) has been developed (EPPO 2010) which provides a generic approach for validating methods of this kind. Validation gives a measure of how a method performs against a range of metrics, enabling end-users to compare different methods and select the most appropriate one. The methods themselves are usually available either in the literature or in international standards (most notably EPPO protocols (www.eppo.int)). The use of recommended protocols (regardless of the availability of validation data) is established and in some cases written into legislation (e.g. Council Directive 98/57/EC on the control of Ralstonia solanacearum). However, writing protocols into legislation may prevent or slow down the adoption of new improved methods due to the work and time required to agree protocols or to rewrite directives as technology develops. While the publication of recommended protocols has clear practical value, following such a protocol does not in itself provide assurance that the method is meeting an approved standard.

In contrast, proficiency testing schemes provide a measure of on-going attainment of an agreed standard, (i.e. is a lab able to meet a pre-defined output) but this approach is not currently established within phytodiagnostic laboratories. Looking to the future, it may be better to invest in the international agreement of the standards of performance that are required for a given test, such that this can be reflected in a panel of proficiency test materials. In this way, any method can be adopted that is capable of meeting the agreed standard (evidenced by published validation data), the ongoing use of which can be monitored by using proficiency testing. This is a more forward looking approach to achieving and maintaining high and consistent standards in a diagnostic laboratory and ensures that new developments can be adopted easily and efficiently.

A more significant barrier to adoption of field based identification methods may be the laboratory diagnosticians who are often relied upon to select and recommend methods for deployment, but may have a vested interest in preventing the use of field methods. Whilst this is understandable, laboratory methods may not provide the best solution in cases where rapid decision making is important to successful outcomes. On the other hand, diagnosticians are frequently the knowledge custodians for pests and pathogens who are relied upon during outbreak situations or when new pathogens and pests are encountered, providing advice on control and containment. Thus there is a fine balance to be struck between deployment of new techniques that provide a more efficient solution, and continuing to maintain specialist knowledge that will be needed in the future.

Acknowledgments We acknowledge funding from Framework 7 of the European Union under grant agreement 245047 for the project Q-detect: Developing quarantine pest detection methods for use by national plant protection organisations (NPPO) and inspection services. NB would in particular like to acknowledge helpful discussions during the course of the project with Maja Ravnikar, Juerg Frey, Andrea Battisti, Francoise Petter, Martin Brandstetter, Maja Zorovich, Cor Schoen, Hugh Mortimer, Jenny Tomlinson, Michael Andreou and Andreas Buhlmann.

References

Andrea L, Joyce AL, Hunt RE, Bernal JS, Vinson SB (2008) Substrate influences mating success and transmission of courtship vibrations for the parasitoid. Cotesia marginiventris 127(1):39–47

Apan A, Held A, Phinn S, Markley J (2004) Detecting sugarcane 'orange rust' disease using EO–1 Hyperion hyperspectral imagery. Int J Remote Sens 25(2):489–498

Augustin S, De Kogel WJ, Donner P, Faccoli M, Lees DC, Marini L, Mori N, Toffolo EP, Quilici S, Roques A, Yart A, Battisti A (2012a) A list of methods to detect quarantine arthropod pests in Europe. OEPP/EPPO Bull 42:93–94

Augustin S, Boonham N, De Kogel WJ, Donner P, Faccoli M, Lees DC, Marini L, Mori N, Petrucco Toffolo E, Quilici S, Roques A, Yart A, Battisti A (2012b) A review of pest surveillance techniques for detecting quarantine pests in Europe. EPPO Bull 42:515–551

Bawden FC (1933) Infra-red photography and plant virus diseases. Nature (Lond) 132:168

Boonham N, Glover R, Tomlinson J, Mumford R (2008) Exploiting generic platform technologies for the detection and identification of plant pathogens. Eur J Plant Pathol 121:355–363

Chamberlain K, Briens M, Jacobs JH, Clark SJ, Pickett JA (2012) Use of honey bees (Apis mellifera L.) to detect the presence of Mediterranean fruit fly (Ceratitis capitata Wiedemann) larvae in Valencia oranges. J Sci Food Agric 92(10):2050–2054

Chesmore D, Schofield J (2010) Acoustic detection of statutory pests in hardwood material. Bull Eur Plant Pathol Organ (EPPO) 40(1):46–51

Chinellato F, Simonato M, Battisti A, Faccoli M, Hardwick S, Suckling DM (2013) Smart-traps combined with molecular on-site detection to monitor Monochamus spp. and associated pine wood nematode. In: Schröder T (ed) Pine wilt disease conference 2013, Braunschweig, pp 23–25. ISSN: 1866-590X

Colebrook FM (1937) The aural detection of the larvae of insects in timber. J Sci Ins 14:119–121

Danks C, Barker I (2000) On-site detection of plant pathogens using lateral-flow devices. EPPO Bull/Bull OEPP 30(3–4):421–426

de Boer JG, Hordijk CA, Posthumus MA, Dicke M (2008) Prey and non-prey arthropods sharing a host plant: effects on induced volatile emission and predator attraction. J Chem Ecol 34(3):281–290

EPPO (2010) PM 7/76 use of EPPO diagnostic protocols. EPPO Bull/Bull OEPP 40:350–352

Franke J, Menz G (2007) Multi-temporal wheat disease detection by multi-spectral remote sensing. Prec Agric 8(3):161–172

Gandelman O, Jackson R, Kiddle G, Tisi L (2011) Loop-mediated amplification accelerated by stem primers. Int J Mol Sci 12(12):9108–9124

James HE, Ebert K, McGonigle R, Reid SM, Boonham N, Tomlinson JA, Hutchings GH, Denyer M, Oura CAL, Dukes JP, King DP (2010) Detection of African swine fever virus by loop-mediated isothermal amplification (LAMP). J Virol Methods 164(1–2):68–74

Jansen RMC, Hofstee JW, Wildt J, Verstappen FWA, Bouwmeester HJ, van Henten EJ (2009) Induced plant volatiles allow sensitive monitoring of plant health status in greenhouses. Plant Signal Behav 4(9):824–829

Jansen RMC, Wildt J, Kappers IF, Bouwmeester HJ, Hofstee JW, van Henten EJ (2011) Detection of diseased plants by analysis of volatile organic compound emission. Ann Rev Phytopathol 49:157–174

Lelong CCD, Burger P, Jubelin G, Roux B, Labbé S, Baret F (2008) Assessment of unmanned aerial vehicles imagery for quantitative monitoring of wheat crop in small plots. Sensors 8(5):3557–3585

Mankin RW, Smith MT, Tropp JM, Atkinson EB, Jong DY (2008) Detection of Anoplophora glabripennis (Coleoptera: Cerambycidae) larvae in different host trees and tissues by automated analyses of sound-impulse frequency and temporal patterns. J Econ Entomol 101(3):838–849

Miresmailli S, Gries R, Gries G, Zamar RH, Isman MB (2010) Herbivore-induced plant volatiles allow detection of Trichoplusia ni (Lepidoptera: Noctuidae) infestation on greenhouse tomato plants. Pest Manag Sci 66(8):916–924

Notomi T, Okayama H, Masubuchi H, Yonekawa T, Watanabe K, Amino N, Hase T (2000) Loop-mediated isothermal amplification of DNA. Nucleic Acids Res 28(12):E63

Potamitis I, Ganchev T, Fakotakis N (2008) Automatic bioacoustic detection of Rhynchophorus Ferrungineus. In: 16th European Signal Processing Conference (EU-SIPCO 2008), Lausanne, Switzerland, 25–29 Aug 2008

Quilici S, Donner P, Battisti A (2012) Surveillance techniques for exotic insect pest detection. Bull OEPP/EPPO 42(1):95–101

Schwarz, Kranz, Sicke (1935) A sound amplifier for the detection of infestations by the house beetle. Deutsche Bauzeitung 20:392–393

Sharp EL, Perry CR, Scharen AL, Boatwright GO, Sands DC, Lautenschlager LF, Yahyaoui CM, Ravet FW (1985) Monitoring cereal rust development with a spectral radiometer. Phytopathology 75(8):936–939

Spinelli F, Cellini A, Vanneste JL, Rodriguez-Estrada MT, Costa G, Savioli S, Harren FJM, Cristescu SM (2012) Emission of volatile compounds by Erwinia amylovora: biological activity in vitro and possible exploitation for bacterial identification. Trees 26:141–152

Steddom K, Jones D, Rush C (2005) A picture is worth a thousand words. APSnet Feature. https://www.apsnet.org/publications/apsnetfeatures/Pages/RemoteSensing.aspx

Tomlinson J, Boonham N (2008) Potential of LAMP for detection of plant pathogens. CAB Rev: Perspectives in Agriculture, Veterinary Science, Nutrition and Natural Resources 3(066):1–7

Tomlinson JA, Barker I, Boonham N (2007) Faster, simpler, more-specific methods for improved molecular detection of Phytophthora ramorum in the field. Appl Environ Microbiol 73(12):4040–4047

Tomlinson JA, Dickinson M, Hobden E, Robinson S, Giltrap PM, Boonham N (2010a) A five-minute DNA extraction method for expedited detection of Phytophthora ramorum following prescreening using Phytophthora spp. lateral flow devices. J Microbiol Methods 81:116–120

Tomlinson JA, Dickinson MJ, Boonham N (2010b) Detection of Botrytis cinerea by loop-mediated isothermal amplification. Lett Appl Microbiol 51:650–657

Tomlinson J, Dickinson M, Boonham N (2010c) Rapid detection of Phytophthora ramorum and P kernoviae by two-minute DNA extraction followed by isothermal amplification, and amplicon detection by generic lateral flow device. Phytopathology 100(2):143–149

Torrance L (1987) Use of enzyme amplification in an ELISA to increase sensitivity of detection of barley yellow dwarf virus in oats and in individual vector aphids. J Virol Methods 15(2):131–138

Wallner WE, Ellis TL (1976) Olfactory detection of gypsy moth pheromone and egg masses by domestic canines. Environ Entomol 5:183–186

Weißbecker B, Schütz S, Klein A, Hummel HE (1997) Analysis of volatiles emitted by potato plants by means of a Colorado beetle electroantennographic detector. Talanta 44(12):2217–2224

Virtual Diagnostic Networks: A Platform for Collaborative Diagnostics

James P. Stack, Jane E. Thomas, Will Baldwin, and Paul J. Verrier

Abstract The successful management of plant diseases and insect infestations depend to some degree on accurate diagnoses delivered in a timeframe that enables effective response. Over the past few decades, many nations have experienced a steady increase in plant pathogen and pest incursions as well as the emergence of new pathogens that threaten plant health. The containment and mitigation of exotic plant pathogens and insect pests are dependent upon early detection and accurate diagnoses also in a timeframe to enable effective response. As we increase our dependence on international trade to address global economic and food security challenges, it is likely that incursions of pathogens and pests will continue, if not accelerate. Climate change, increasing standards of living, and globalized market systems will put increasing demands and pressures on the plant systems that provide us with food, feed, fiber, and medicines. It is appropriate to ask: do we have a rational plant biosecurity strategy and adequate plant biosecurity infrastructure to protect plant systems from the pathogens and pests that threaten plant health?

Keywords Plant health • Diagnostic networks • Plant disease diagnosis • Plant pest identification • Plant disease management • Pest management

---

J.P. Stack (✉) • W. Baldwin
Department of Plant Pathology, Biosecurity Research Institute, Kansas State University, 66506 Manhattan, KS, USA
e-mail: jstack@ksu.edu

J.E. Thomas
Department of Plant Pathology, National Institute of Agricultural Botany (NIAB), Huntingdon Road, Cambridge, CB3 0LE, UK

P.J. Verrier
Department of IT and Computing, National Institute of Agricultural Botany (NIAB), Huntingdon Road, Cambridge, CB3 0LE, UK

M.L. Gullino and P.J.M. Bonants (eds.), *Detection and Diagnostics of Plant Pathogens*, Plant Pathology in the 21st Century 5, DOI 10.1007/978-94-017-9020-8_10,

Plant Health at Risk

Natural and agricultural plant systems will face significant biosecurity challenges over the course of this century. Climate change is predicted to result in the geographic redistribution of plants, plant pathogens, and vectors and may contribute to the evolution and emergence of new pathogens, novel pathogen-vector associations and new diseases. Global trade in plants and plant materials is predicted to result in the geographic redistribution of plants, plant pathogens, and vectors (Dehnen-Schmutz et al. 2010; Shaw and Osborne 2011) and may contribute to the evolution and emergence of new pathogens, novel pathogen-vector associations and new diseases.

The emergence of hybrid Phytophthora species has been reported as a consequence of the global trade in nursery and landscape plants (Brasier 2001). Of great concern is that the host range and aggressiveness of the hybrid Phytophthora species were not predicted based on the parental phenotypes. Novel virulence patterns in plant pathogenic bacteria have been attributed to horizontal gene transfer (HGT). The global movement of plants may provide more opportunity for HGT among plant pathogenic bacteria creating challenges for plant health in general, and plant diagnostics, in particular. The ability of cryptic satellites to alter the host range of Gemini viruses may present new challenges to plant health and confound our ability to diagnose and mitigate these new virus diseases. Viable plant seeds have been recovered from melting permafrost as a consequence of warming trends in northern Russia (Yashinaa et al. 2012). If the plants are viable after being frozen for 10,000 years, plant-associated microbes, including pathogens, may also be viable. The vulnerability of current plant genotypes to ancient pathogen genotypes is unknown.

The global trade networks that link plant production systems to food and ornamental plant markets (MacLeod et al. 2010; Pautasso et al. 2010) have connected nations with well-developed plant biosecurity infrastructure to nations with poor and ineffective plant biosecurity infrastructure. From 2000 to 2009, exports from developing nations and imports into developing nations as a percentage of the global mean have exceeded that for developed nations by over 30 % (url). There is ample evidence that when we move plants and plant products, we also move pathogens, arthropod pests, and invasive plant species (Brasier 2008; Britton 2004). Invasive species including, plant pathogens and insect pests have been distributed in food aid shipped to countries during times of critical food shortages. Although one could argue that to adequately protect plant systems we need to limit plant trade (Brasier 2008), there are no reasonable scenarios for the future that do not include the trade of plants and plant products fully integrated into a global economy. Trade will be necessary to meet the food security needs of some nations and essential to the economic wellbeing of many nations. The United Nations Food and Agriculture Organization (FAO) projects that, even with our best aid and development efforts, some nations will remain food insecure and will require food assistance to alleviate malnutrition and emergency food aid to avert starvation

(FAO 2009). Plant-based foods are the primary source of nutrition for most of the world. Consequently, for the foreseeable future, we will continue to ship plants and plant products around the world. Over the previous decade, the World Food Program has responded to approximately 30 food emergencies each year to avert starvation.

The Need for Robust Plant Diagnostic Systems

Plant diagnostic systems are critical to attaining and maintaining plant health and to protecting plant trade (Stack and Fletcher 2007). From seed to silo and farm to fork, plants are vulnerable to a very large array of diseases pre and postharvest. With respect to trade, the mere presence of a pathogen can stop shipments whether or not disease was manifest. Being able to accurately identify a pathogen and diagnose a disease is essential to ensure proper disease mitigation measures are implemented and to ensure that correct trade decisions are made. With respect to sensitivity, specificity, and robustness, the stringency of diagnostic technologies and protocols vary across the continuum from pre-infection detection to post-outbreak monitoring.

Prevention

As with human health, prevention is a very effective strategy for maintaining plant health. This is true for recurrent, emergent, and exotic plant pathogens. With respect to exotic plant pathogens, pre-border, point-of-origin inspections of plants and plant products are increasing in importance. Often times these inspections are symptom-based examinations which are limited in scope and prone to error if infections are latent. Over the past decade, there has been a rapid increase in the development and validation of advanced, field deployable diagnostic technologies. These field deployable diagnostic technologies range from inexpensive lateral-flow style devices for detecting protein or nucleic acid signals to very expensive, and often cumbersome, real time polymerase chain reaction thermal cyclers. More recently, field-friendly isothermal amplification devices have come into use. High specificity, excellent sensitivity, and fast run times in the field are among the advantages of many of these new diagnostic tools.

In addition to point-of-origin inspections of plant-based commodities prior to export, declaring plant production areas free of specific pathogens is also growing in use as a practice to facilitate trade. If an organism has never been reported in an area or if an area survey has been conducted and determined to be negative for a specific pathogen, then plants and plant products can be exported from that area with few restrictions.

Declaring an area free of a specific pathogen and pre-export inspections of plants and plant products depend to a great extent on validated surveillance and sampling designs and on accurate, robust diagnostic protocols. The consequences of false negatives and false positives can be devastating to both plant health and to trade. Diagnostic protocols with a low propensity for false negatives are important to prevent incursions and protect plant health while diagnostic protocols with a low propensity for false positives are important to preclude unnecessary trade interruptions.

Early Detection and Accurate Diagnoses

Once a pathogen has become established or disease has been initiated, early detection is critical to minimizing the ultimate impacts. As time lapses from pathogen establishment or disease outbreak, the severity of the disease and the distance that the pathogen spreads from the outbreak or establishment site often increases. Consequently, early detection is the key to keeping both the severity of the disease and the spread of the pathogen to a minimum. Accurate diagnostics with low false negatives (minimize introductions) and low false positives (minimize trade interruptions) are essential to assess the effectiveness of the mitigation and containment measures taken. Our knowledge about pathogen populations is increasing and with that come changes in the boundaries that define individual taxa. We must constantly develop new diagnostic tools to account for our ever changing understanding of plant pathogens (Tinivella et al. 2008).

Response and Recovery

Rapid and effective response is based upon the assumption of a correct identification of the pathogen. The deployment of appropriate disease mitigation measures is dependent upon that same identification. Response and mitigation based on an incorrect identification may either be ineffective or exacerbate the outbreak thus delaying or perhaps even precluding recovery. Accurate diagnoses are most often the critical step in minimizing the negative impacts resulting from disease outbreaks. It is not just at the onset of an outbreak or even during the rapid expansion phase of an epidemic that diagnostics are important. Diagnostic support is critical throughout the entire response and recovery effort. Diagnostic support is critical not just for the first detection, but throughout the entire response and recovery effort from the onset of an outbreak through the rapid expansion phase of an epidemic, to the resolution phase of the event that diagnostics are important. Because of that the diagnostic capability (i.e., technology and experience) and capacity (i.e., sample surge support) become important assets of any plant diagnostic system.

Often after an incursion or a disease outbreak, a survey to delimit the area affected is conducted. One effect of most delimiting surveys is a rapid and substantial increase in the number of samples to be diagnosed. This can very quickly overwhelm most plant diagnostic laboratories or clinics. Processing large numbers of samples quickly may require high throughput technologies and labor; sometimes beyond the capabilities of individual plant diagnostic labs. Another approach to surge support is a coordinated response by several diagnostic laboratories each with limited throughput capability but combined, capable of processing a very large number of samples in a short period of time. This same approach can be effective when attempting to establish areas of freedom for a specific pathogen which may also require processing a very large number of samples.

Virtual Plant Diagnostic Networks

We have developed a global trade network for the large-scale distribution of plants and plant products including foods, ornamental and landscape plants, timber, and other plant-based goods. The productivity and sustainability of plant systems around the world have been negatively impacted by the unintended movement of pathogens and pests associated with the distribution of those plants and plant products. To protect the plant systems upon which public health and economic wellbeing depend, we need a global network of plant diagnostic infrastructure to facilitate the detection, diagnosis, and management of recurrent, emergent, and introduced plant pathogens.

Many nations have identified increased trade as a priority to address poor economies. However, trade increases the risk to plant health as a direct result of the movement of unwanted plant pathogens and pests. The International Plant Protection Convention (IPPC) requires that signatory nations abide by strict Sanitary and Phytosanitary (SPS) standards to minimize the movement of plant pathogens and pests across national boundaries (MacLeod et al. 2010). However, the systems to ensure that only pathogen-free plants are put into global distribution chains are overwhelmed by the massive volume of plants that are shipped over great distances in short periods of time. The practice of shipping plants between continents in less time than the latent period for many diseases calls into question the effectiveness of interception protocols that are based on symptoms. Although well intended and fairly well implemented, SPS procedures are only partially effective resulting in regular introductions of plant pathogens into new environments.

Plant Diagnostic Networks

The concept for a plant health network is not new. The American Phytopathological Society (APS) proposed such a network almost 100 years ago (Campbell et al. 1999). In 1918, the leadership of APS recognized that a nationally coordinated response to plant health emergencies would be more effective than an uncoordinated set of independent efforts. That concept is just as valid today. In 2002, the Land Grant University (LGU) system in the U.S. partnered with the United States Department of Agriculture to create the National Plant Diagnostic Network (NPDN) to increase the capability and capacity of plant diagnostics throughout the U.S. (Stack et al. 2006; Stack 2010). The plant diagnostic laboratories in most states were housed in LGUs but limited in scope and resources. The tripartite mission of NPDN is to promote the early detection of plant pathogens and pests by training first detectors, to ensure accurate and rapid diagnostics by investment in diagnostic infrastructure and the training of diagnosticians in each state, and to facilitate timely and secure communications among diagnostic laboratories and regulatory agencies (Stack et al. 2006; Stack and Baldwin 2008). The U.S. Agency for International Development partnered with a few LGUs to create the International Plant Diagnostic Network (IPDN) to extend this concept to nations with very limited diagnostic resources (Miller et al. 2009). In-country training programs regarding the art, science, and practice of plant diagnostics, including modern diagnostic technologies have been conducted (Miller et al. 2010). The Global Plant Clinic (now called PlantWise) was created by CABI to bring plant diagnostic support to rural communities in low income nations (Bentley et al. 2003; Boa 2007). PlantWise has offered on-site clinics in rural settings in South America, Asia, and Africa and where internet connectivity exists, PlantWise offers on-line digital diagnostic support from the United Kingdom.

The European Union Framework programs have funded multinational efforts to create a virtual biosecurity research and diagnostic network for Europe (Gullino et al. 2008). The concept is to link researchers and diagnosticians across Europe into a mutually beneficial network to enhance plant diagnostic capabilities and perhaps eventually providing first detectors in the field with access to information necessary to ensure early detection of emerging or introduced plant pathogens and pests. Among the challenges are linking nations with different languages, different institutional cultures, different regulatory frameworks, and encouraging cooperation among trading partners.

Plant Health Australia is working to establish a national plant diagnostic network (http://www.planthealthaustralia.com.au/national-plant-biosecurity-diagnostic-network) with many of the same goals as the other networks discussed. Scientists in Australia have created an on-line plant biosecurity toolbox (http://old.padil.gov.au/pbt/) to provide access to diagnostic resources and have already deployed a digital diagnostics system (http://www.padil.gov.au/Rmd) for Australia that now extends that capability into Southern Asia.

Magnitude of the Challenge for Plant Diagnostics

Unlike public health systems with responsibility for one host, plant diagnosticians have responsibility for a vast diversity of host plant species often spanning four levels of taxonomic complexity (e.g., varieties, species, genera, families). Each one of these plant species has a vast diversity of pathogens that cause an array of diseases, each with its own set of symptoms, dynamics, host specificity, and ever-changing systematics and nomenclature of the pathogens. Remaining current with the changing taxonomies, technologies and protocols necessary to diagnose across kingdoms (e.g., Eubacteria, Fungi, Chromalveolata) is challenging. Equally diverse are the number and nature of clientele served by plant diagnostic laboratories ranging from homeowners with sick turf and ailing ornamentals to public botanical gardens with hundreds of plant species from around the world to large-scale corporate farming operations with a potentially devastating new disease of an important food crop. This diversity of clientele plays out in the value of a diagnostic test and the speed with which an accurate diagnosis needs to be rendered.

Plant diagnostic laboratories at air and sea ports of entry have enormous challenges with respect to the number of shipments, the number of samples to be processed, and the very short timeframes within which diagnoses must be rendered. When each airplane and ship arrives in port with a diverse array of plant species to be inspected, there are numerous trucks outside the inspection station waiting for the shipments to be cleared for distribution so that the plants can be transported to retail centers in good condition. Some plants and plant products have short shelf lives and consequently, any delays can decrease the value of the plants.

Attributes of a Plant Diagnostic Network

The coordination of diagnostic efforts during plant health emergencies is one of the primary benefits to a plant diagnostic network. The benefits of coordination include, the timely distribution of accurate information needed for response efforts, identification and access to expertise and diagnostic supplies, and a rapid delimitation of the area affected. When there is a sample surge and diagnostic kits or reagents are limited in supply, the diagnostic network can facilitate either moving samples to where the kits and reagents are in supply or facilitating the movement of kits and reagents to the diagnostic labs experiencing a surge of samples. When protocols fail due to either the emergence of new biotypes of pathogens or a failure of reagents, that information can be quickly distributed throughout the network to avoid the costly and damaging problems that result from false negatives and false positives. Knowing where to access expertise and knowing the capabilities and capacities of laboratories with respect to diagnostic technologies and experience is invaluable during sample surge and during delimiting surveys. The large number of host species and even larger number of pathogen species makes unlikely that any one diagnostician can be an expert in them all. One of the greatest attributes of a diagnostic network is access to expertise, no matter where that expertise resides in relation to the need.

Data sharing and information security are critical considerations for a plant diagnostic network (Burrows et al. 2009; Stack and Baldwin 2008). The distribution of trade-sensitive and response-sensitive information must be kept secure. Consequently, all the stakeholders of plant diagnostic information must be considered before choosing the communications technologies and before developing communications protocols. Those technologies and protocols must be practiced to ensure data security.

Concept for a Virtual International Plant Diagnostic Network (VIPDN)

A few national and regional plant diagnostic networks are functioning well (Miller et al. 2010; Stack 2010). Additional national and regional plant diagnostic networks are being designed and developed. To complement those national and regional networks, we propose a virtual international plant diagnostic network (VIPDN). The mission and function of a VIPDN would be: (1) to provide a mechanism to facilitate the exchange of non-trade sensitive diagnostic information and resources, (2) to provide a vehicle for cooperation among plant diagnosticians around the world, and (3) to provide a directory of plant diagnostic laboratories to facilitate interaction with the global research community.

The confidentiality of plant diagnostic data is necessary to ensure the uninterrupted trade of plants and plant products. The release of preliminary, unconfirmed plant diagnostic data can damage the industries that support plant systems and the economies that are dependent upon plant systems. Within the concept of VIPDN, each national and regional network would retain ownership of its diagnostic data and maintain the information management systems to limit access to those data. However, there is a lot of plant diagnostic information and expertise in laboratories around the world that would be of great value to the international plant diagnostic community at large. A VIPDN could be the platform for sharing diagnostic resources such as images of symptoms, primer and probe sequences, the advantages and limitations of specific diagnostic protocols, technological expertise such as isothermal amplification, etc. A VIPDN could also be the platform for raising awareness regarding training opportunities for diagnosticians with respect to specific taxa and/or technologies. It would likely require the coordination of an international scientific society (e.g., the International Society of Plant Pathology, www.isppweb.org/) and the support of an institution to develop and host the website. The benefits to the advancement of plant diagnostics and plant protection would far outweigh the costs (Fletcher 2008). Cooperation and collaboration in plant diagnostics on an international scale makes sense; after all, my indigenous are your exotics and your indigenous are my exotics.

Conclusion

The biological and climatological challenges to plant health underscore the need for a renewed commitment to the science and practice of plant diagnostics. Central to that need are well-trained and highly skilled plant diagnosticians working in well-funded, well-equipped, and technologically sophisticated plant diagnostic clinics. These needs will likely increase over the course of this century. Our inability to meet the demands for early detection and accurate plant diagnostics may lead to failures in our plant protection systems and subsequent declines in the health of agricultural and landscape plant systems. Professional development programs for plant diagnosticians and added investment in diagnostic infrastructure are essential. National, regional and international plant diagnostic networks will greatly increase the capacity and capability for plant diagnostics and thus improve plant health globally.

Acknowledgements This project was funded in part from the European Union Seventh Framework Programme (FP7/2007-2013) under Grant Agreement n° 261752. Contribution number 14-298-B from the Kansas Agricultural Experiment Station.

References

Bentley JW, Boa E, Van Mele P, Almanza J, Vasquez D, Eguino S (2003) Going public: a new extension method. Int J Agric Sust 1:108–123

Boa E (2007) Plant healthcare for poor farmers: an introduction to the work of the global plant clinic. Online. APSnet Features. doi:10.1094/APSnetFeatures-2007-1007

Brasier CM (2001) Rapid evolution of introduced plant pathogens via interspecific hybridization. Bioscience 51(2):123–133

Brasier CM (2008) The biosecurity threat to the UK and global environment from international trade in plants. Plant Pathol 57:792–808

Britton KO (2004) Controlling biological pollution. In: Britton K (ed) Biological pollution, an emerging global menace. American Phytopathological Society, St. Paul

Burrows M, Franc G, Rush C, Blunt T, Ito D, Kinzer K, Olson J, O'Mara J, Price J, Tande C, Ziems A, Stack J (2009) Occurrence of viruses in wheat in the great plains region, 2008. Online. Plant Health Progress. doi:10.1094/PHP-2009-0706-01-RS

Campbell CL, Peterson PD, Griffith CS (1999) The formative years of plant pathology in the United States. APS Press, St. Paul, p 448

Dehnen-Schmutz K, Holdenrieder O, Jeger M, Pautasso M (2010) Structural change in the international horticultural industry: some implications for plant health. Sci Hortic 125 (2010):1–15

Fletcher J (2008) The case for international cooperation as a strategy to achieve crop biosecurity. In: Gullino ML, Fletcher J, Gamliel A, Stack JP (eds) Crop biosecurity: assuring our global food supply. Springer, Dordrecht, pp 111–120

Food and Agriculture Organization (2009) How to feed the world in 2050. FAO, Rome

Gullino ML, Gamliel A, Fletcher J (2008) Evaluation of critical issues and identification of research and investment needs in the field of crop biosecurity in Europe. In: Gullino ML, Fletcher J, Gamliel A, Stack PJ (eds) Crop biosecurity: assuring our global food supply. Springer, Dordrecht, 148 pp

MacLeod A, Pautasso M, Jeger MJ, Haines-Young R (2010) Evolution of the international regulation of plant pests and challenges for future plant health. Food Sec 2:49–70. doi:10.1007/s12571-010-0054-7

Miller SA, Beed F, Harmon CL (2009) Plant disease diagnostic capabilities and networks. Annu Rev Phytopathol 47:15–38

Miller SA, Kinyua ZM, Beed F, Harmon CL, Xin J, Vergot P, Momol T, Gilbertson R, Garcia L (2010) The International Plant Diagnostic Network (IPDN) in Africa: improving capacity for diagnosing diseases of banana and other African crops. Acta Hort 879:341–348

Pautasso M, Dehnen-Schmutz K, Holdenrieder O, Pietravalle S, Salama N, Jeger MJ, Lange E, Hehl-Lange S (2010) Plant health and global change – some implications for landscape management 2010. Biol Rev 85:729–755. doi:10.1111/j.1469-185X.2010.00123.x

Shaw MW, Osborne TM (2011) Geographic distribution of plant pathogens in response to climate change. Plant Pathol 60:31–43

Stack JP (2010) Diagnostic networks for plant biosecurity. In: Hardwick N Gullino ML (eds) Knowledge and technology transfer for plant pathology. Springer, Dordrecht, 123 p

Stack JP, Baldwin W (2008) The need for secure communications networks and global connectivity. In: Gullino ML, Fletcher J, Gamliel A, Stack PJ (eds) Crop biosecurity: assuring our global food supply. Springer, Dordrecht, 148 pp

Stack JP, Fletcher J (2007) Plant biosecurity infrastructure for disease surveillance and diagnostics, pp 95–106. In: Institute of Medicine. 2007. Global infectious disease surveillance and detection: assessing the challenges – finding the solutions. Workshop summary. The National Academies Press, Washington, DC, 263 p

Stack J, Cardwell K, Hammerschmidt R, Byrne J, Loria R, Snover-Clift K, Baldwin W, Wisler G, Beck H, Bostock R, Thomas C, Luke E (2006) The national plant diagnostic network. Plant Dis 90:128–136

Tinivella F, Gullino ML, Stack JP (2008) The need for diagnostic tools and infrastructure, pp 63–71. In: Gullino ML, Fletcher J, Gamliel A, Stack PJ (eds) Crop biosecurity: assuring our global food supply, Springer, Dordrecht, 148 pp

Yashinaa S, Gubinb S, Maksimovichb S, Yashinaa A, Gakhovaa E, Gilichinskyb D (2012) Regeneration of whole fertile plants from 30,000-y-old fruit tissue buried in Siberian permafrost. Proc Natl Acad Sci U S A 109(10):4008–4013

Development and Implementation of Rapid and Specific Detection Techniques for Seed-Borne Pathogens of Leafy Vegetable Crops

Maria Lodovica Gullino, Giovanna Gilardi, Giuseppe Ortu, and Angelo Garibaldi

Abstract The usage of commercial seeds produced in few facilities increases the potential for introducing and disseminating several pathogens into new production areas. Fast and sensitive diagnostic tools are necessary to screen seeds and transplants. Identifying the source of inoculum is of critical importance for effective disease management. Due to the role of seed transmission, detection methods are also important for the production of pathogen-free seeds and for their certification. Pathogen's detection on seeds is difficult because in most cases infected seeds can be asymptomatic, making visual detection difficult or impossible. Moreover, infected seeds may be present in a limited percent, and not evenly distributed within a lot. The development of the Polymerase Chain Reaction opened new perspectives in plant pathogen's detection and today the most used assays are based on this technique or on new techniques developed in order to improve its sensitivity. This chapter focuses on recent developments for the quick detections of pathogens of leafy vegetables of recent introduction. During the past 15 years, several new diseases were introduced, in Italy as well as in many other countries, in this production sector through infected seeds. The molecular methods developed for the detection of several *formae speciales* of *Fusarium oxysporum*, *Verticillium dahliae* and several causal agents of foliar diseases of leafy vegetables are described and critically discussed.

Keywords PCR • Molecular diagnostics • Soilborne pathogens • Seedborne pathogens

---

M.L. Gullino (✉)
AGROINNOVA, University of Torino, Grugliasco, Torino, Italy
e-mail: marialodovica.gullino@unito.it

G. Gilardi • G. Ortu • A. Garibaldi
Center of Competence Agroinnova, University of Torino,
Via Leonardo da Vinci 44, 10095 Grugliasco, Italy

M.L. Gullino and P.J.M. Bonants (eds.), *Detection and Diagnostics of Plant Pathogens*, Plant Pathology in the 21st Century 5, DOI 10.1007/978-94-017-9020-8_11,

Introduction

With recent increase in usage of commercial seeds produced in few facilities, there is greater potential for introducing and disseminating pathogens into new production areas. Fast and sensitive diagnostic tools are needed so that primary inoculum vectors, such as seeds and transplants, can be screened promptly. Pathogen's detection on seeds can be a difficult task because in most cases infected seeds can be asymptomatic, making visual detection difficult or impossible. Moreover, infected seeds may be present in a limited percent, and non uniformly distributed within a lot. Different detection assays exist for different seed-borne pathogens but only a few respect the minimum requirements for adequate seed tests. Traditionally, seed assays have been developed based on (i) visual examination: in some cases infected seeds can present symptoms as discoloration or shriveling making it possible a visual detection of the infection. This assay was used in the past for seed-borne disease in purple seed of soybean (Murakishi 1951) or peanuts (Randall-Schadel et al. 2001); (ii) selective media: pathogen is grown on artificial media, this assay can be done by plating directly surface sterilized seed or by plating wash seed suspension. Subsequently, the pathogen isolated can be identified. Unfortunately on the seed also saprophytic microorganisms able to grow in the rich media can be present, thus making difficult the isolation of the pathogens (Walcott 2003). (iii) serological techniques: based on specific interaction between antibody (mono or polyclonal) and specific antigen on the surface of plant pathogens (Hapton et al. 1990). Several kinds of these tests are know but the most important is ELISA (enzyme linked immunosorbent assay) (McLaughlin and Chen 1990). Afterwards, the development of the Polymerase Chain Reaction (Mullis and Faloona 1987) opened new perspectives in plant pathogen's detection and the most used assays are based on this technique (Henson and French 1993). Using pathogen DNA as a target and two specific primers designed on specific sequence it is possible to determine the infection without pathogen isolation. PCR-based methods show several merits, first of all the short time to obtain the results (2–3 h), second the sensitivity (low copy of DNA can be detectable) and, finally, the easy results interpretation (Taylor et al. 2001). However, the major negative aspect of the PCR technique is the possibility to obtain the false positive from non viable pathogens (Agarwal 2006). To overcome this aspect a new technique has been developed called BIO-PCR in which the pathogen is cultured on synthetic media before the amplification. This method was used for detecting several bacterial pathogens and in a few cases also for fungi (Smith et al. 1996).

During the past 15 years, several new diseases of leafy vegetables (lettuce, wild and cultivated rocket, lamb's lettuce, cichory, endive, basil, spinach) were introduced in Italian cultivations through infected seeds (Garibaldi and Gullino 2010). Some of them were reported for the first time in Europe or worldwide. Identifying the source of inoculum is of critical importance for effective disease management. Due to the role of seed transmission, detection methods are also important for the production of pathogen-free seeds and for their certification.

Among diseases, Fusarium wilts have been recently observed in Italy on lettuce (*Lactuca sativa*) (Garibaldi et al. 2002, 2004a), wild (*Diplotaxis* spp.) and cultivated rocket (*Eruca sativa*) (Garibaldi et al. 2003, 2004c), lamb's lettuce (*Valerianella olitoria*) (Garibaldi et al. 2004b), cichory (*Cichorium intybus*) (Garibaldi et al. 2011a) and endive (*Cichorium endivia*) (Garibaldi et al. 2009). Also several species of *Alternaria*, for which there is evidence of being seed transmitted, are reported on leafy vegetables. *Alternaria cichorii* is since longtime reported on lettuce, endive and scarola, while *Alternaria japonica* has been recently detected on both wild and cultivated rocket (Garibaldi et al. 2011b). *Alternaria alternata* is the causal agent of leaf spot on basil (Garibaldi et al. 2011c).

Among other pathogens recently observed on leafy vegetables, *Verticillium dahliae*, reported on lettuce (Garibaldi et al. 2007), cichory (Ciccarese 1987) and spinach (du Toit et al. 2005), *Plectosphaerella cucumerina* on wild rocket (Garibaldi et al. 2012), *Phoma valerianellae*, reported since 1966 on lamb's lettuce (Garibaldi 1966), recently reemerged as a major problem in many Italian farms (Pellegrino et al. 2010). Since the conventional pathogen detection techniques may lack the sensitivity required to detect seed-borne pathogens, molecular techniques have been developed which permit a quick and reliable identification, also increasing the detection thresholds of several pathogens. This chapter will focus on recent developments for the quick detections of pathogens of leafy vegetables of recent introduction.

Fusariumoxy oxysporum

The search for molecular techniques has been particularly intensive and effective in the case of several *formae speciales* of *Fusarium oxysporum* (Lievens et al. 2007, 2008, 2012), also due to the importance of Fusarium wilts on many economically relevant crops. Several molecular approaches, such as sequence typing and use of molecular markers, have been exploited in order to differentiate *formae speciales* and races within *the F. oxysporum* species complex (Lievens et al. 2012). The detection threshold of *F. oxysporum* in seeds and propagation material could be increased by using molecular techniques, such as the polymerase chain reaction (PCR) as already shown in the case of Fusarium wilt of basil (Garibaldi et al. 1997; Chiocchetti et al. 2001; Pasquali et al. 2006) and lettuce (Pasquali et al. 2008).

Molecular methods have been developed to provide fast and unequivocal identification of *F. oxysporum* f. sp. *basilici*. A nested-PCR-based method (Chiocchetti et al. 2001) allowed sensitive detection directly from wilted plants and seed lots without the isolation of the pathogen. The method permitted to detect 32 conidia/100 seeds and required 4 h. With this method, DNA was extracted only from propagules present on the external surface of the seeds. The detection of *F. oxysporum* f. sp. *basilici* was further improved by developing a real-time PCR method, which resulted very sensitive and reproducible (Pasquali et al. 2006). The real-time PCR assay was able to detect 24 ± 10 CFU/100 seeds, permitted to detect the pathogen on internally and externally infected seeds and could be performed in 1.5 h. Its cost, of about 1 Euro per reaction, is twice that of nested-PCR. However, considering that only half of the time is required for real-time PCR, compared with nested-PCR, the cost is levelled (Pasquali et al. 2006). Moreover, the possibility of performing direct extraction on seed, avoiding the washing procedure requested by nested-PCR, is a further reason for adopting real-time PCR for the detection of *F. oxysporum* f. sp. basilici. The real-time PCR method, tested on Italian isolates of the pathogen, due to the clonal origin of the *forma specialis* (Katan et al. 1996) and to the presence of the sequence used for designing the real-time PCR primers in the populations of *F. oxysporum* f. sp. *basilici* from South Africa (Swart and Van Niekerk, 2003) and Australia (Summerell et al. 2006), should also be valid for isolates of *F. oxysporum* f. sp. *basilici* isolated from other areas (Pasquali et al. 2006). The major disadvantage of this last method is that it is unable, like all molecular methods based on DNA amplification, to distinguish viable from dead propagules. This limitation can be overcome using a RT–PCR technique. Since the mRNA is degraded quickly in dead cells, the detection of mRNA can considered a correct indicator of cell viability (Sheridan et al. 1998). RNA extracted from seeds can be reverse transcribed using the reverse transcriptase and the cDNA obtained can be used as target for any other PCR-based method. RT-PCR has been used to detect viable populations of *Mycosphaerella graminicola* in wheat (Guo et al. 2005) and *Oidium neolycopersici* in tomato (Matsuda et al. 2005). Some seed producers sell now certified Fusarium-free basil seeds based on the use of the real-time PCR method for detection of *F. oxysporum* f. sp. *basilici*. Molecular methods have been used to distinguish between nonpathogenic and pathogenic forms of *F. oxysporum*. Mitochondrial haplotype analysis and sequence analysis of the mitochondrial small subunit (mtSSU), translocation elongation factor 1- α (EF-1 α) gene, and nuclear ribosomal (r) DNA intergenic spacer (IGS) region have been used to determine diversity among *formae speciales* other than *lactucae* (Appel and Gordon 1995; O'Donnell et al. 1998; Skovgaard et al. 2001). Mbofung et al. (2007) evaluated these methods for their utility in determining phylogenetic relationships among *F. oxysporum* f. sp. *lactucae* isolates, nonpathogenic *F. oxysporum* isolates, and selected *formae speciales*. They found that mtSSU sequence analysis did not differentiate *lactucae* isolates from 13 other *F. oxysporum* isolates. Analysis of sequences provided some resolution, grouping the *lactucae* isolates with seven *formae speciales*. However, the IGS region contained the most sequence variation and provided the best resolution of the *lactucae* isolates as distinct taxa. With IGS region analysis, lactucae race 1 isolates resolved as a monophyletic group along with three other *formae speciales* of *F. oxysporum*. In all analyses, *lactucae* race 2 isolates segregated as a separate lineage phylogenetically distinct and distantly related to *lactucae* race 1 isolates. Comparable findings were reported by Fujinaga et al. (2005) using only partial IGS sequences. Fujinaga et al. also examined *lactucae* race 3 and found that it was genetically distinct from both race 1 and 2, which suggests an additional lineage of *F. oxysporum* f. sp. *lactucae*.

Phylogenetic analysis based on IGS and EF-1 α regions was also used for the characterization of *F. oxysporum* spp. on *Cichorium intybus*, with the description of a new *forma specialis* called *F. oxysporum* f. sp. *cichorii* (Poli et al. 2012).

A sequence-specific amplified polymorphism (SSAP) technique developed to study the clonality of *F. oxysporum* f. sp. *lactucae* VCG 0300 permitted discrimination among race 1 and race 2 strains. Polymorphisms within Italian and US race 1 isolates were detected but no relation was found between genetic variability and geographic distribution of strains (Pasquali et al. 2008). This technique confirmed the worldwide clonality of VCG 0300 race 1. Moreover, it can be used for reliable strain discrimination and as a tool to investigate transposition behaviour of retroelements in the genome of the pathogen (Pasquali et al. 2008). A nested polymerase chain reaction-based (nPCR) assay developed for rapid detection of *F. oxysporum* f. sp. *lactucae* in lettuce seeds by Mbofung and Pryor (2010) permitted to detect the pathogen in seed lots with infestation rate as low as 0.1 %.

Verticillium dahliae

A quantitative real time polymerase chain reaction (qPCR) assay was recently optimized and used for the detection and quantification of *V. dahliae* in spinach seeds, resulting quite reliable and sensitive, permitting to assess values up to 1.3 % of seed infection (Duressa et al. 2011). However, *V. longisporum* has been described as a hybrid species that presents several genomic regions in common with *V dahliae* (Inderbitzin et al. 2011). *V. longisporum* is a crucifer pathogen never described in diseases associated with spinach, but in any case the external presence of this pathogen can result into a false positive; then it is important to exclude the presence of this pathogen from the samples.

Alternaria

PCR-based methods were used for detecting *Alternaia radicina* infections on carrot (Pryor and Gilbertson 2001). Based on the RAPD analysis on different *Alternaria* species, the authors designed two specific primers able to distinguish *A. radicina* from the other *Alternaria* species. In 2002 Konstantinova et al. on the basis of their PCR-RLFP analysis, designed three specific oligos able to amplify *A. radicina*, *A. dauci* and *A. alternata* always from carrot seeds. More recently a standard PCR and a Real time PCR were used in order to identify seed contamination from *A. brassicae* on cabbage and radish seeds (Guillemette et al. 2004).

Real time PCR was used to detect *A. brassicicola* in *Arabidopsis* tissues (Brouwer et al. 2003). However, the specificity of this assay could be criticized due to the lack of other *Alternaria* species in the experimental set-up. In 2004, Gachon and Saindrenan designed real time primers always for *A. brassicicola* based on cutinase A gene, but also in this case the specificity of this primer was compromised for the low presence of this gene sequences in the database. Taqman real time reverse-trascriptase PCR technique was developed in order to obtain an *Alternaria* spp. viable quantification on 110 commercial foodstuff by Pavon et al. (2012) In this study specific *Alternaria* spp. primers have been designed on ITS1 and ITS2 region achieving a 1 CFU/g as detection limit. Although the current real challenge in pathogen detection on seeds is to obtain a viable detection and quantification, only few studies use a RNA isolation together with the PCR on plant tissue and nobody applied it to the seed-borne pathogens.

Other Pathogens

A real-time PCR assay based on internal transcribed spacer (ITS) sequence was developed to detect *Plectosphaerella cucumerina* on *Diplotaxis tenuifolia* seeds (Webb et al. 2014). Such assay is highly specific and does not amplify five non target species of *Plectosphaerella*, as well as other 18 fungal species, thus improving the assay previously designed by Atkins et al. (2003) which, by showing cross reaction with all *Plectosphaerella* species tested, could not be used for specific detection of *P. cucumerina*. The new assay developed by Webb et al. is able to detect DNA concentrations of 10 fg. However, it should be further improved to increase its sensitivity for seed testing (2014).

In the case of lamb's lettuce, a PCR method was designed for detection of *Phoma valerianellae* in seeds: variation within the internal transcriber spacer (ITS1, 5.8S sequences and ITS2) region of the rDNA (ITS) was used to characterize the pathogen and to design specific primers within the ITS region (Pellegrino et al. 2010). The assay developed is highly specific and can be used on a large scale to distinguish *P. valerianellae* from other *Phoma* species (Pellegrino et al. 2010). Unfortunately this primer can not be used on real-time PCR because the amplicon size (496 bp) is too large for this technique.

Conclusions

Seed health is a prerequisite for sustainable agriculture. Seed testing permits to avoid further field treatments with a much higher environmental impact. Moreover, seed detection assays is important for excluding that dangerous pathogens are introduced or spread by seeds. Seed health tests should be sensitive, specific, reliable, rapid and not too expensive. Traditional seed-borne detection assays are currently used, but in the last years, PCR-based methods have achieved better results for their specificity, sensitivity, rapidity, ease of implementation, interpretation and applicability. Since leafy vegetable crops have a growing economic importance, also due to the increasing consumption of ready-to-eat salads, seed producers can invest in more sophisticated and innovative diagnostic tools. The different assays described for the different pathogens are, in general, enough sensitive, with detection limits of about 1 infected seed out of 100. The PCR assays last one day in comparison with at least 2 weeks for conventional plate or blotter assays. Therefore, PCR-methods open interesting and new approaches for the seed-borne detection. However, these new technique do not provide information about the pathogen vitality. Furthermore, these methods are still more expensive compared to the traditional seed health assays that also do not require sophisticated and expensive equipments. In conclusion PCR-based method are relatively expensive but are fast and sensitive. In the future, advances in technologies will permit to process larger number of samples, thus reducing operational costs.

Acknowledgements The results reported have received funding from the European Union's Seventh Framework Programme (FP7/2007–2013) under grants agreement n° 261752 PlantFoodSec – Plant and Food Biosecurity, Network of Excellence and 311875 "TESTA – Seed health: development of seed treatment methods, evidence for seed transmission and assessment of seed health".

References

Agarwal VK (2006) Seed health. International Book Distributing, Lucknow

Appel DJ, Gordon TR (1995) Intraspecific variation within populations of *Fusarium oxysporum* based on RFLP analysis of the intergenic spacer region of the rDNA. Exp Mycol 19:120–128

Atkins SD, Clark IM, Sosnowska D, Hirsch PR, Kerry BR (2003) Detection and quantification of *Plectosphaerella cucumerina*, a potential biological control agent of potato cyst nematodes, by using conventional PCR, real-time PCR, selective media, and baiting. Appl Environ Microbiol 69:4788–4793

Brouwer M, Lievens B, van Hemelrijk W, van den Ackerveken G, Cammue BPA, Thomma BPHJ (2003) Quantification of disease progression of several microbial pathogens on *Arabidopsis thaliana* using real-time fluorescence PCR. FEMS 228:241–248

Chiocchetti A, Sciaudone I, Durando F, Garibaldi A, Migheli Q (2001) PCR detection of *Fusarium oxysporum* f. sp. *basilici* on basil. Plant Dis 85:607–611

Ciccarese F (1987) Severe outbreaks of verticillium wilt on *Cichorium intybus* and *Brassica rapa* and pathogenic variations among isolates of *Verticillium dahliae*. Plant Dis 71:1144–1145

Du Toit LJ, Derie ML, Hernandez-Perez P (2005) Verticillium wilt in spinach seed production. Plant Dis 89:4–11

Duressa D, Rauscher G, Koike ST, Mou B, Hayes RJ, Marutchachalam K, Subbarao KV, Klosterman SJ (2011) A real-time PCR assay for detection and quantification of *Verticillium dahliae* in spinach seed. Phytopathology 102:443–451

Fujinaga M, Ogiso H, Shinohara H, Tsushima S, Nishimura N, Togowa M, Saito H, Nosue M (2005) Phylogenetic relationships between the lettuce root rot pathogen *Fusarium oxysporum* f. sp. *lactucae* race 1, 2, and 3 based on the sequence of the intergenic spacer region of its ribosomal DNA. J Gen Plant Pathol 71:402–407

Gachon C, Saindrenan P (2004) Real-time PCR monitoring of fungal development in *Arabidopsis thaliana* infected by *Alternaria brassicicola* and *Botrytis cinerea*. Plant Physiol Biochem 42:367–371

Garibaldi A (1966) Due nuove malattie della valerianella. In: Proceedings 1st Congress Mediterranean Phytopathological Union, Bari-Napoli, Italy, pp 153–156

Garibaldi A, Gullino ML (2010) Emerging soilborne diseases of horticultural crops and new trends in their management. Acta Hortic 883:37–46

Garibaldi A, Gullino ML, Minuto G (1997) Diseases of basil and their management. Plant Dis 81:124–132

Garibaldi A, Gilardi G, Gullino ML (2002) First report of *Fusarium oxyspor*um on lettuce in Europe. Plant Dis 86:1052

Garibaldi A, Gilardi G, Gullino ML (2003) First report of *Fusarium oxyspor*um on *Eruca vesicaria* and *Diplotaxis* sp. in Europe. Plant Dis 87:201

Garibaldi A, Gilardi G, Gullino ML (2004a) Seed transmission of *Fusarium oxysporum* f. sp. *lactucae*. Phytoparasitica 32:61–65

Garibaldi A, Gilardi G, Gullino M (2004b) First report of *Fusarium oxysporum* causing vascular wilt of Lamb's lettuce (*Valerianella olitoria*) in Italy. Plant Dis 88:83

Garibaldi A, Gilardi G, Pasquali M, Keiji S, Gullino ML (2004c) Seed transmission of *Fusarium oxysporum* of *Eruca vesicaria* and *Diplotaxis muralis*. J Plant Dis Prot 111:345–350

Garibaldi A, Gilardi G, Gullino ML (2007) First report of Verticillium wilt caused by *Verticillium dahliae* on lettuce in Italy. Plant Dis 91:770

Garibaldi A, Gilardi G, Troisi M, Gullino ML (2009) First report of Fusarium wilt of endive (*Cichorium endivia*) caused by *Fusarium oxysporum*. Plant Dis 10:1,078.3

Garibaldi A, Gilardi G, Poli A, Gullino ML (2011a) First report of Fusarium wilt of chicory (*Cichorium intybus*) caused by *Fusarium oxysporum* in Italy. Plant Dis 95:496

Garibaldi A, Gilardi G, Bertoldo C, Gullino ML (2011b) First report of a leaf spot of sweet basil (*Ocimum basilicum*) caused by *Alternaria alternata* in Italy. J Plant Pathol 93(supp 4s):4.71

Garibaldi A, Gilardi G, Bertoldo C, Gullino ML (2011c) First report of leaf spot of wild (*Diplotaxis tenuifolia*) and cultivated (*Eruca vesicaria*) rocket caused by *Alternaria japonica* in Italy. Plant Dis 95:1316

Garibaldi A, Gilardi G, Ortu G, Gullino ML (2012) First report of *Plectosphaerella cucumerina* on wild rocket (*Diplotaxis tenuifolia*) in greenhouses in Italy. Plant Dis 96:1825

Guillemette T, Iacomi-Vasilescu B, Simoneau P (2004) Conventional and real-time PCR-based assay for detecting pathogenic *Alternaria brassicae* in cruciferous seed. Plant Dis 88:490–496

Guo JR, Schnieder F, Beyer M, Verreet JA (2005) Rapid detection of *Mycosphaerella graminicola* in wheat using reverse transcription-PCR assay. J Phytopathol 153:674–679

Hampton R, Ball E, De Boer S (eds) (1990) Serological methods for detection and identification of viral and bacterial plant pathogens: a laboratory manual. APS Press, St. Paul

Henson JM, French R (1993) The polymerase chain reaction and plant disease diagnosis. Annu Rev Phytopathol 31:81–109

Inderbitzin P, Davis RM, Bostock RM, Subbarao KV (2011) The ascomycete *Verticillium longisporum* is a hybrid and a plant pathogen with an expanded host range. PLoS One 6:18260

Katan T, Gamliel A, Katan J (1996) Vegetative compatibility of *Fusarium oxysporum* from sweet basil in Israel. Plant Pathol 45:656–661

Konstantinova P, Bonants PJM, van Gent-Pelzer MPE, van der Zouwen P, van der Bulk R (2002) Development of specific primer for detection and identification of *Alternaria* spp. in carrot material by PCR and comparison with blotter and plating assay. Mycol Res 106:23–33

Lievens B, Claes L, Vakalounakis DJ, Vanachter ACRC (2007) A robust identification and detection assay to discriminate the cucumber pathogens *Fusarium oxysporum* f. sp. *cucumerinum* and f. sp. *radicis-cucumerinum*. Environ Microbiol 9:2145–2161

Lievens B, Rep M, Thomma BPHJ (2008) Recent developments in the molecular discrimination of *formae speciales* of *Fusarium oxysporum*. Pest Manag Sci 64:781–788

Lievens B, Hanssen IM, Rep M (2012) Recent developments in the detection and identification of *formae speciales* and races of *Fusarium oxysporum*: from pathogenicity testing to molecular diagnosis. In: Gullino ML, Katan J, Garibaldi A (eds) Fusarium wilts of greenhouse crops. APS Press, St. Paul, pp 47–55

Matsuda Y, Sameshima T, Moriura N, Inoue K, Nonomura T, Kakutani K, Nishimura H, Kusakari S, Tamamatsu S, Toyoda H (2005) Identification of individual powdery mildew fungi infecting leaves and direct detection of gene expression of single conidium by polymerase chain reaction. Phytopathology 95:1137–1143

Mbofung GC, Pryor BM (2010) A PCR-based assay for detection of *Fusarium oxysporum* f. sp. *lactucae* in lettuce seed. Plant Dis 94:860–866

Mbofung GY, Hong SG, Pryor BM (2007) Phylogeny of *Fusarium oxysporum* f. sp. *lactucae* inferred from mitochondrial small subunit, elongation factor 1-α, and nuclear ribosomal intergenic spacer sequence data. Phytopathology 97:87–98

McLaughlin RJ, Chen TA (1990) ELISA methods for plant pathogenic prokaryotes. In: Hampton R, Ball E, DeBoer S (eds) Serological methods for detection and identification of viral and bacterial plant pathogens. APS Press, St. Paul, pp 197–204

Mullis KB, Faloona FA (1987) Specific synthesis of DNA in vitro via a polymerase catalyzed chain-reaction. Method Enzymol 155:335–350

Murakishi HH (1951) Purple seed stain of soybean. Phytopathology 41:305–318

O'Donnell K, Kistler HC, Cigelink E, Ploetz RC (1998) Multiple evolutionary origins of the fungus causing Panama disease of banana: concordant evidence from nuclear and mitochondrial gene genealogies. Proc Natl Acad Sci U S A 95:2044–2049

Pasquali M, Piatti P, Gullino ML, Garibaldi A (2006) Development of a real-time polymerase chain reaction for the detection of *Fusarium oxysporum* f. sp. *basilici* from basil seed and roots. J Phytopathol 154:632–636

Pasquali M, Saravanakumar D, Gullino ML, Garibaldi A (2008) Sequence specific amplified polymorphism (SSAP) technique to analyse VCG 0300 of *Fusarium oxysporum* f. sp. *lactucae*. J Plant Pathol 90:527–535

Pavón MÁ, Gonzáles I, Martín R, García T (2012) A real-time reverse-transcriptase PCR technique for detection and quantification of viable Alternaria spp. in foodstuff. Food Control 28:286–294

Pellegrino C, Gilardi G, Gullino ML, Garibaldi A (2010) Detection of *Phoma valerianellae* in lamb's lettuce seeds. Phytoparasitica 38:159–165

Poli A, Gilardi G, Spadaro D, Gullino ML, Garibaldi A (2012) Molecular characterization of *Fusarium oxysporum* f. sp. *cichorii* pathogenic on cichory (*Cichorium intybus*). Phytoparasitica 40:383–391

Pryor BM, Gilbertson RL (2001) A PCR assay for detection of Alternaria radicina on carrot seed. Plant Dis 85:18–23

Randall-Schadel BL, Bailey JE, Beute MK (2001) Seed transmission of *Cylindrocladium parasiticum* in peanut. Plant Dis 85:362–370

Sheridan GEC, Masters CI, Shallcross JA, Mackey BM (1998) Detection of mRNA by reverse transcription-PCR as an indicator of viability in *Escherichia coli* cells. Appl Environ Microb 4:1313–1318

Skovgaard K, Nirenberg HI, O'Donnell K, Rosendahl S (2001) Evolution of *Fusarium oxysporum* f. sp. *vasinfectum* races inferred from multigene genealogies. Phytopathol 91:1231–1237

Smith OP, Peterson GL, Bec RJ, Schaad NW, Bonde M (1996) Development of a PCR-based method for identification of *Tilletia indica*, casual agent of Karnal bunt of wheat. Phytopathology 86:115–122

Summerell BA, Gunn LV, Bullock S, Tesoriero LT, Burgess LW (2006) Vascular wilt of basil in Australia. Australas Plant Pathol 35:65–67

Swart L, van Niekerk JM (2003) First record of *Fusarium oxysporum* f. sp. *basilici* occurring on sweet basil in South Africa. Australas Plant Pathol 32:125–126

Taylor E, Bates J, Kenyon D, Maccaferri M, Thomas J (2001) Modern molecular methods for characterization and diagnosis of seed-borne fungal pathogens. J Plant Pathol 83:75–81

Walcott RR (2003) Detection of seedborne pathogens. HortTechnology 13:40–47

Webb K, Gilardi G, Matic S, Gullino ML, Garibaldi A (2014) Development of a rapid *in planta* detection method for identifying *Plectosphaerella cucumerina* on wild rocket by real-time PCR. Eur J Plant Pathol (accepted, in press)

Diagnosis of Plant Pathogens and Implications for Plant Quarantine: A Risk Assessment Perspective

Vittorio Rossi, Thierry Candresse, Michael J. Jeger, Charles Manceau, Gregor Urek, and Giuseppe Stancanelli

Abstract The Scientific Panel on Plant Health (PLH) of the European Food Safety Authority was established in 2006 as the reference EU body for risk assessment in the plant health area in the frame of phytosanitary measures to prevent the introduction and further spread of organisms harmful (pests) to plants or plants products. Proper diagnosis and detection are critical for the categorization and assessment of pests that are or may qualify as quarantine pests, and for appropriate application of phytosanitary measures. In carrying out its responsibilities, the PLH Panel is frequently faced by the changes in taxonomy and nomenclature: these may often lead to a re-evaluation of previous pest records, concerns with published literature,

V. Rossi (✉)
Università Cattolica del Sacro Cuore, Via E. Parmense 84, 29100 Piacenza, Italy
e-mail: vittorio.rossi@uinicatt.it

T. Candresse
INRA and University of Bordeaux Virology Team, UMR 1332 Biologie du Fruit et Pathologie, CS20032, 33882 Villenave d'Ornon Cedex, France
e-mail: tc@bordeaux.inra.fr

M.J. Jeger
Centre for Environmental Policy, Imperial College London, Silwood Park campus, Ascot SL5 7PY, UK
e-mail: m.jeger@imperial.ac.uk

C. Manceau
Anses, French Agency for Food, Environmental and Occupational Health and Safety, 7 rue Jean Dixméras, 49044 Angers, France
e-mail: charles.manceau@anses.fr

G. Urek
Kmetijski inštitut Slovenije, Hacquetova ulica 17, 1000 Ljubljana, Slovenia
e-mail: gregor.urek@kis.si

G. Stancanelli
European Food Safety Authority, Animal and Plant Health Unit, Via Carlo Magno 1A, 43126 Parma, Italy
e-mail: giuseppe.stancanelli@efsa.europa.eu

M.L. Gullino and P.J.M. Bonants (eds.), *Detection and Diagnostics of Plant Pathogens*, Plant Pathology in the 21st Century 5, DOI 10.1007/978-94-017-9020-8_12,

and in some cases with the increased use of molecular taxonomic and identification tools. Pathogens that were previously considered strains/varieties of a single species, in some cases have now been elevated to specific status. In other cases pathogens previously recognised as species complexes have now been partitioned into named species, using molecular criteria rather than traditional methods. The problems that can arise are illustrated in this Chapter by examples representative of the main pathogen taxa.

Keywords Pest risk assessment • Taxonomic and nomenclature changes • Phytophthora diseases • Fungal diseases • Bacterial diseases • Viruses and viroids • Nematodes

The Plant Health Legislation

The International Regulation Framework

The international regulatory framework in plant health is set by the Sanitary and Phytosanitary (SPS) Agreement of the World Trade Organisation (WTO). Based on the SPS agreement, countries have the right to take sanitary and phytosanitary measures, provided that these are based on scientific principles and are not maintained without sufficient scientific evidence (WTO-SPS 1995). To ensure that they are globally harmonized, phytosanitary measures shall be based on international standards, guidelines or recommendations developed by the FAO Secretariat of the International Plant Protection Convention (IPPC), in cooperation with the regional plant protection organizations. The "scientific principles" of phytosanitary measures are based on risk assessment, which includes the evaluation of the likelihood of entry, establishment or spread of a pest or disease within the territory of an importing country and of the associated potential biological and economic consequences. The IPPC has produced a set of standards, the International Standards for Phytosanitary Measures (ISPM) (www.ippc.int/core-activities/standards-setting/ispms). The regional plant protection organisations – for Europe, the European and Mediterranean Plant Protection Organisation (EPPO) – have produced more detailed regional guidelines and decision support schemes for pest risk assessment (www.eppo.int/).

The European Approach

The Council Directive 2000/29/EC enforces the protective measures to prevent the introduction and further spread into Member States of the European Union of organisms harmful to plants or plants products. The plant pests and pathogens banned from the EU countries on any or on specified plants and plant products, as well as their specific phytosanitary requirements, are listed in the annexes of the Directive. Separate lists are provided for plant pests and diseases banned from the whole EU and for those banned from particular EU areas, known as protected zones. These lists cover all EU plant quarantine insects, mites, nematodes, bacteria, phytoplasma, chromista, fungi, viruses, virus-like organisms and parasitic plants, generally identified at species level, with a few exceptions, for example, for diseases of unknown etiology or for groups of organisms with comparable biology as in the case of viruses vectored by *Bemisia tabaci*.

With a view of basing risk assessment on scientific principles and keeping it separate from risk management and decision making, the Regulation (EC) No 178/2002, established the European Food Safety Authority (EFSA) as the reference EU risk assessment body to provide independent scientific advice and communication in food safety, animal health and welfare, and plant health. The EFSA Scientific Panel on Plant Health (PLH) (www.efsa.europa.eu/en/panels/plh.htm) was later established in 2006, with the Commission Regulation (EC) 575/2006 of 7 April 2006 amending Regulation (EC) No 178/2002, as the EU reference body for risk assessment in the plant health area, recognising the protection of plant health as an essential factor in the security of the food chain (Schans et al. 2008). Production of pest risk assessments, identification of pest risk reduction options and evaluation of their effect on the level of pest risk, as well as the review and evaluation of pest risk analyses made by third parties, represent the core outputs of the scientific advice produced by the Panel, together with the development of new methods and guidance for pest risk assessment (Jeger et al. 2012).

Diagnosis of Plant Pathogens and Plant Health Regulation

Diagnostic Needs

In the context of IPPC standards, diagnosis is defined as "the process of detection and identification of a pest" (ISPM No. 27). A pest is defined as "any species, strain or biotype of plant, animal or pathogenic agent injurious to plants or plant products" (ISPM No. 2). Pest identification is a key element within the process of pest risk analysis for the purpose of preparing phytosanitary regulations by NPPOs (ISPM No. 2). Pest risk analysis (PRA) consists of three stages: (i) initiating the process for analyzing risk; (ii) assessing pest risk; and (iii) managing pest risk. Initiating the process involves identification of pests that may qualify as quarantine pests or of pathways for which the risk analysis is needed; either of these two starting points can involve pests already present in the PRA area but not widely distributed and under official control, as well as pests absent from the PRA area, since both are covered by the quarantine pest definition. Pest risk assessment determines whether each pest identified as such, or associated with a pathway, is a quarantine pest, characterized in terms of likelihood of entry, establishment, spread and economic importance. Pest risk management involves developing, evaluating, comparing and selecting options for reducing the risk.

Proper detection and/or identification are also critical for the appropriate application of phytosanitary measures as, for example, illustrated in a range of international documents: Requirements for the establishment of pest-free areas (ISPM No. 4); Guidelines for surveillance (ISPM No. 6); Export certification system (ISPM No. 7); Guidelines for pest eradication programmes (ISPM No. 9); and Guidelines for a phytosanitary import regulatory system (ISPM No 20). In addition, diagnostic procedures are needed for determination of pest status in an area (ISPM No. 8), pest reporting (ISPM No. 17), and the diagnosis of pests in imported consignments (ISPM No. 13).

In general, the provision of reliable pest records is a vital component of a number of activities covered under the IPPC and by the principles noted in ISPM No. 1: Principles of plant quarantine as related to international trade, and the international standards for phytosanitary measures that have been developed from them. The basic information needed in a pest record consists of the current scientific name of the organism including, as appropriate, sub-specific terms (strain, biotype, etc.). Reliability of pest identification has been defined by ISPM No. 8; concerning technical identification of the organism, the following are listed from most reliable to least reliable: discriminating biochemical or molecular diagnosis (if available); specimen or culture maintained in official collection, taxonomic description by specialist; specimen in general collection; description and photo; visual description only; method of identification not known.

NPPOs have produced diagnostic protocols for regulated pests in order to adequately fulfill their responsibilities according to Article IV of the IPPC (1997); in particular, regarding surveillance, import inspections and export certification. In response to the need for regional harmonization, several RPPOs have developed a number of regional diagnostic standards, including those developed by the European and Mediterranean Plant Protection Organisation (http://www.eppo.int/STANDARDS/standards.htm). This underlines the need for international harmonization of diagnostic protocols within the framework of the IPPC. IPPC approved the formation of a Technical Panel on Diagnostic Protocols for that purpose.

ISPM No. 27 provides guidance on the structure and content of the diagnostic protocols for regulated pests. The protocols describe procedures and methods for the official diagnosis of regulated pests that are relevant for international trade. They provide the minimum requirements for reliable diagnosis of a regulated pest, which include information relevant for diagnosis, the pest taxonomic position, and the methods to detect and identify it. Diagnostic protocols also provide flexibility to ensure that methods are appropriate for use across the full range of circumstances. The methods included in diagnostic protocols are selected on the basis of their sensitivity, specificity and reproducibility, and information related to these elements is provided for each of the methods. Detailed information and guidance for the detection of pests is provided on, for example, signs and/or symptoms associated with the pest, illustrations (where appropriate), developmental stages of the pest, and methods for detecting the pest in a commodity, as well as methods for extracting, recovering and collecting the pests from plants. Information and guidance for the identification of pests includes detailed information on morphological and morphometric methods, methods based on biological properties, and methods based on biochemical and molecular properties of the pest. Protocols are subject to review and amendment to take into account new developments in pest diagnosis.

Problems for Risk Assessment Posed by Taxonomic Changes

In carrying out its responsibilities for risk assessment in plant health, the EFSA PLH Panel in many cases is faced by the changes in taxonomy and nomenclature that have taken place since the Council Directive in 2000. This may often lead to a re-evaluation of pest records held in EU Member States, concerns with published literature, and, in some cases with the increased use of molecular taxonomic and identification tools, doubts as to whether a regulated organism is actually the one of concern. Pathogens that were previously considered strains/varieties of a single species, have now in some cases been elevated to specific status. In other cases pathogens recognised as species complexes have been partitioned into recently-named species, using molecular rather than traditional methods. Molecular methods now offer a more finely-tuned assessment of the genetic variation present at the subspecific level: an aspect of increasing importance in risk assessment. The problems that can arise are illustrated below by examples representative of the main pathogen taxa that have been the subject of Opinions published by the PLH Panel in the last 5 years.

Case-Studies

Phytophthora *Diseases*

Diseases caused by the oomycete genus *Phytophthora* have long been important in agriculture, horticulture, forestry and native vegetation. In the last two decades the number of known *Phytophthora* species has increased markedly largely due to the increasing global trade in plants for planting or for sale. This had led to an increasing need for improved diagnosis, techniques for detection, and interception in ensuring compliance with plant health regulation. Here we describe the importance of diagnostic methods using as examples: *Phytophthora ramorum*, a newly emerging and previously unknown pathogen (EFSA Panel on Plant Health 2011b); and *Phytophthora fragariae* causing red core disease in cultivated strawberry (EFSA Panel on Plant Health in preparation).

Phytophthora ramorum

The exotic nature of the pathogen discovered as causal agent of sudden oak death (SOD) in North America and of a new *Phytophthora* species from European nurseries was based on the observations that the species had never previously been described in either continent (Werres et al. 2001; Rizzo et al. 2002), the distribution was either geographically limited (California), or clearly associated with the nursery trade (Europe), the high susceptibility of hosts in natural settings (Rizzo et al. 2002); and the presence of different mating types in the two continents (Brasier 2003; Ivors et al. 2004).

Ivors et al. (2004) showed that the genetic structure of California forest populations was indicative of a strong genetic bottleneck, but was distinct from European nursery isolates. Ivors et al. (2006) identified three genetically distinct lineages in *P. ramorum*, with the distribution of lineages being: NA1 in North American nurseries and forests, NA2 in North American nurseries, and EU1 in European nurseries. The evidence indicated that introduction in the nursery trade must have occurred multiple times, both in North America and in Europe.

A linkage between nursery populations of the pathogen and those present in California forests or parks in Europe was apparent (Garbelotto and Rizzo 2005). Mascheretti et al. (2008) provided strong genetic evidence that nursery populations of the pathogen were ancestral to California forest populations. Spread within North America and Europe also occurred through the movement of infected ornamental plants (Goss et al. 2009; Prospero et al. 2009). The presence of site-specific genotypes in Belgian nurseries (Vercauteren et al. 2010) and in California forests (Mascheretti et al. 2008, 2009) suggested an on-going micro-evolutionary process. It now appears that nurseries no longer contribute genotypes to established forest populations in California and Oregon (Prospero et al. 2007; Mascheretti et al. 2008).

In comparing phenotypes of different lineages (Brasier and Kirk 2004; Elliott et al. 2009), differences in mating type, growth rate, and virulence were identified. The three known lineages are extremely similar in broad ecological and virulence traits (Elliott et al. 2009); however, the co-occurrence of NA1/NA2 and EU1 lineages would present a potential threat because of the possibility of sexual reproduction (Boutet et al. 2010). Despite most evidence indicating a lack of sexual reproduction (Ivors et al. 2006; Vercauteren et al. 2010, 2011); Boutet et al. (2010) reported it was possible and that progeny displayed a range of virulences with the potential for further adaptation. The area of origin of *P. ramorum* is unknown (Hansen 2010). A recent report of *P. lateralis* in a *Chamaecyparis* forest in Taiwan (Brasier et al. 2010) suggests that, as the two species are phylogenetically related and share common features, they are likely to share a common region of origin. *Chamaecyparis* is present both in Taiwan and Japan, and both *P. ramorum* and *P. lateralis* could have originated from one of these two countries. The recent report of *P. lateralis* causing root and aerial infections on *Chamaecyparis lawsoniana* hedgerow trees in Brittany (Robin et al. 2011), the first report of the pathogen outside nurseries apart from Oregon and California (Kliejunas 2010), suggests that *P. lateralis* poses a new plant health threat to Europe.

Phytophthora fragariae

Phytophthora fragariae was first described as the organism causing red core of strawberry (Hickman 1940). It was subsequently split into var. *fragariae*, which causes red core of strawberry, and var. *rubi*, which causes root rot of red raspberry (Wilcox et al. 1993). In 1997, var. *rubi* was elevated to a new species *P. rubi* on the basis of a gene flow analysis (Man in't Veld 2007). The original description of Hickman was then re-applied only to the *Phytophthora* causing red core of strawberry (EFSA Panel on Plant Health 2014).

P. fragariae and *P. rubi* are closely related to one another and were both assigned to a clade within *Phytophthora*, arbitrarily named Clade 7a, among which *P. cambivora* appears to be the nearest to *P. fragariae* and *P. rubi*. All isolates of *P. fragariae*, so far tested, have very similar whole protein electrophoretic patterns distinct from *P. rubi* or *P. cambivora* (Duncan et al. 1991) and all isolates with this protein profile produced typical red core symptoms on strawberry plants. The same isolates caused only minor symptoms on raspberry plants (*Rubus idaeus* L.). The exact reverse was the case with isolates of *P. rubi*. Hence, *P. fragariae* is the one species causing red core disease of strawberry (Duncan et al. 1991).

The symptoms of red core disease make diagnosis relatively straightforward and unambiguous. The main roots rot from the tips upwards, and the lack of lateral roots gives affected roots a characteristic appearance, with a red discolouration of their steles. Thick-walled oospores are produced in abundance in and around the stele. In summer, the pathogen becomes quiescent, depending for survival upon oospores left in rotted roots.

Cultural characteristics have been well described. Cultures of *P. fragariae* on French bean agar form large amounts of evenly-fluffy aerial mycelium but grow more slowly than other species in Clade 7a, even *P. rubi*. *P. fragariae* produces fairly large non-papillate sporangia; however, most if not all of the developmental features of sporangia and sporangiospores can be found in other Clade 7a species. *P. fragariae* does not form oospores in culture, in contrast to *P. rubi*.

Molecular markers have been used to discriminate *P. fragariae* from closely related species. The ITS regions of ribosomal DNA is a good target for routine identification of *P. fragariae*, although used alone it does not distinguish *P. fragariae* from *P. rubi*. A combination of the cytochrome c oxidase gene, a mitochondrial gene, and ITS sequence (Robideau et al. 2011) should separate *P. fragariae* and *P. rubi*.

Susceptible *Fragaria* spp. such as *F. vesca var. alpina* have been used as bait plants to detect *P. fragariae* in the field and in soil samples; these have been used to demonstrate survival of *P. fragariae* in infested sites for at least 10 years after a diseased strawberry crop (Newton et al. 2010). The soil bait test has been adapted to test commercial stocks, as validated in an EU Cost programme (Duncan 2001).

Detection methods using specific primers based on the ITS sequences of rDNA (Cooke et al. 2000) have been used to detect *P. fragariae* and *P. rubi* in two-round and single-round PCR. The root tip bait test was modified and combined with PCR testing to yield a range of testing protocols for red core that were validated by nine EU-based laboratories (Duncan 2001). Rapid PCR-product detection protocols have subsequently been developed (Bonants et al. 2004).

Fungal Diseases

In the two case-studies presented here – ray blight disease of chrysanthemum and needle blight of conifers – the recent changes in the taxonomy of the causal agents – mainly due to the use of new, molecular approaches – make the nomenclature used in the current legislation based on Council Directive 2000/29/EC difficult to interpret and pose problems for application of the Directive. For chrysanthemum ray blight, the taxonomic rearrangement of the *Stagonosporopsis* complex implies that information on geographic distribution of the regulated organism (i.e., *Didymella ligulicola*) should be treated with caution. For pine needle blight, the taxonomic distinction between *Dothistroma septosporum* (which is currently regulated as the former name of the teleomorph *Scirrhia pini*) and the new species *Dothistroma pini* raise the problem that *D. pini*, not being listed in Council Directive 2000/29/EC, is not regulated and that no phytosanitary measures are available to prevent its introduction into or spread within the EU territory.

Stagonosporopsis chrysanthemi

Ray blight is an important disease of chrysanthemum which affects all plant parts, with flowers and cuttings being particularly susceptible (Baker et al. 1949). The causal organism of the chrysanthemum ray blight disease is regulated in the Council Directive 2000/29/EC as *Didymella ligulicola* (Baker, Dimock and Davis) v. Arx, on plants of *Dendranthema* (DC) Des Moul. intended for planting, other than seeds.

Following a request from the European Commission, the EFSA PLH Panel was asked to deliver a scientific opinion on the pest risk for the European Union territory. Delivery of this opinion faced the considerable taxonomic confusion occurred in the past over the causal agent of the ray blight disease of chrysanthemum (EFSA Panel on Plant Health 2013a). Two varieties had been identified in the past within the regulated species *D. ligulicola*: (i) *D. ligulicola* var. *ligulicola* (syn. *Phoma ligulicola* var. *ligulicola*), and (ii) *D. ligulicola* var. *inoxydabilis* (syn. *P. ligulicola* var. *inoxydabilis*) (Van der Aa et al. 1990). Both varieties exist in Europe, and probably elsewhere, with the latter occurring on various wild and cultivated Asteraceae (Van der Aa et al. 1990). The epithet *ligulicola* was introduced by Baker et al. (1949) after describing the teleomorph in *Mycosphaerella*, to avoid confusion with *Sphaerella chrysanthemi* (Tassi 1900) described in Italy. Also the names *Mycosphaerella chrysanthemi* (Tomilin 1979) and *Didymella chrysanthemi* (basionym *Sphaerella chrysanthemi*) have been widely used erroneously, based on the supposed identity between *Mycosphaerella ligulicola*, described in the USA, and *S. chrysanthemi*, reported in Europe. Walker and Baker (1983) demonstrated that the American species is distinct and it did not reach Europe until the 1960s. *D. ligulicola* var. *ligulicola* was recently recognised in the genus *Stagonosporopsis* (Aveskamp et al. 2010) and the name *Stagonosporopsis ligulicola* var. *ligulicola* was proposed, with *Mycosphaerella ligulicola* (Baker et al. 1949) as the basionym. However, after transferring the species to *Stagonosporopsis*, the older epithet, i.e. *chrysanthemi*, based on *Ascochyta chrysanthemi* (Stevens 1907), was preferred over *ligulicola*, thus the combination *Stagonosporopsis chrysanthemi* is the current name (Vaghefi et al. 2012). *S. chrysanthemi* is the causal agent of ray blight of chrysanthemum in the USA and Europe; the pathogen has been reported to affect only cultivars of *Chrysanthemum* × *morifolium* (Van der Aa et al. 1990). *D. ligulicola* var. *inoxydabilis* has been redescribed as *Stagonosporopsis inoxydabilis* (Vaghefi et al. 2012); *S. inoxydabilis* infects *Tanacetum cinerariifolium* but only in Europe (Van der Aa et al. 1990; Vagethi et al. 2012), *T. parthenium*, *Matricaria* sp., and *Zinnia elegans* (Van der Aa et al. 1990). In addition, a third species, *Stagonosporopsis tanaceti*, was recognised as the causal agent of ray blight disease of *Tanacetum cinerariifolium* in Australia and, following artificial inoculation, other Asteraceae species (Pethybridge et al. 2008a; Vagheti et al. 2012).

Since all the above-mentioned *Stagonosporopsis* species cause ray blight symptoms, reliable diagnosis of the organism included into the current EU legislation, *S. chrysanthemi* is only possible by laboratory testing based on the cultural and morphological characters of the pathogen as well as on its positive reaction to NaOH (i.e. production of red pigment β) (Van der Aa et al. 1990; de Gruyter et al. 2002; Vaghefi et al. 2012). However, since identification of *Phoma*-like fungi based on morphological features alone requires special expertise, no standard diagnostic protocol exists for the identification of *S. chrysanthemi*, and there are no specific detection molecular methods available for *S. chrysanthemi* – the PCR method described by Pethybridge et al. (2004) is not specific for the detection of *S. chrysanthemi*, as the primers used cannot differentiate between the three *Stagonosporopsis* species included in the *D. ligulicola* (syn. *P. ligulicola*) complex – molecular sequencing would be necessary for correct identification (Aveskamp et al. 2008, 2009, 2010; de Gruyter et al. 2009, 2010, 2013).

As a consequence of the above-mentioned taxonomic confusion in the past over the causal agent of the chrysanthemum ray blight disease, the information on geographic distribution of the organism should be treated with caution due to the uncertainty about the correspondence between the organism isolated in the past from chrysanthemum plants showing ray blight-like symptoms and the causal agent of ray blight under its current taxonomic status (i.e. *S. chrysanthemi*).

Dothistroma septosporum and *D. pini*

The red band needle blight disease affects a range of conifer species, but *Pinus* spp. are the most common hosts. Affected needles typically develop yellow and tan spots and bands, which soon turn red; as the disease progresses, the ends of the needles turn reddish-brown whilst the needle base remains green. The yellow-tan spots rapidly turn into brown, reddish-brown bands which can remain visible after the needle has died and dropped from the tree. Needle blight symptoms are similar to those caused by other pine needle pathogens (e.g., *Mycosphaerella dearnessii*, *Cercoseptoria pini-densiflorae*, etc.). The causal agent of the disease is listed in Council Directive 2000/29/EC as *Scirrhia pini*; according to Annex II, Part A, Section II of the Directive, introduction into and spread within all EU Member States of *S. pini* shall be banned if it is present on plants of *Pinus* L., intended for planting, other than seeds.

There has been much taxonomic confusion over the causal agent of the needle blight disease (EFSA Panel on Plant Health 2013b). The current EU plant health legislation is for *Scirrhia pini*, but this teleomorph has been renamed *Mycosphaerella pini*, and the anamorph of *M. pini* is now known as *Dothistroma septosporum*. In addition, another species, *Dothistroma pini* (teleomorph unknown), has recently been identified as a second causal agent of needle blight. *D. septosporum* and *D. pini* are closely related and cause the same disease (*Dothistroma* needle blight) predominantly on pine hosts.

The two species are similar in ecology and morphology, and relatively large variation occurs between isolates of the same species (Barnes et al. 2008). Since the two species were only differentiated in 2004 using molecular tools (Barnes et al. 2004), it may be assumed that information available in the literature on *D. pini* prior to 2004, or where molecular analysis to confirm the species has not been undertaken, would now be considered as referring to the anamorph of *M. pini*, *D. septosporum*, because in the past the name *D. pini* has been a synonym for *D. septosporum*. The known distribution of *D. pini* is limited. However, in countries where both species occur, they are found under similar environmental conditions, and in the case of Hungary they have even been found on the same needle (Barnes et al. 2011). Therefore, it is possible that a number of the earlier reports of *D. septosporum* were actually *D. pini*.

Reliable identification of *D. septosporum* and *D. pini* are possible only by laboratory testing based on their cultural, morphological, and molecular characteristics. However, the two species are indistinguishable morphologically unless molecular tools are used. Barnes et al. (2004) used DNA from portions of the nuclear ribosomal internal transcribed spacer (ITS), β-tubulin and elongation factor 1-α to separate the two species. Groenewald et al. (2007) used degenerate primers to amplify portions of mating type genes (*MAT1-1-1* and *MAT1-2*) and chromosome walking to obtain the full-length genes in both species. Thus, the mating type-specific primers could distinguish between the morphologically similar *D. pini* and *D. septosporum*, and between the different mating types using conventional PCR.

In 2010, Ioos et al. developed both conventional and real-time PCR methods to distinguish between *D. septosporum, D. pini* and *Lecanosticta acicola* (syn. *Scirrhia acicola*, teleomorph *Mycosphaerella dearnessii*), a closely related conifer pathogen that causes the brown spot needle blight disease. The Joint Genome Institute (USA) has recently sequenced the *D. septosporum* genome, which will allow further molecular identification techniques to be developed (DOE JGI 2012).

Bacterial Diseases

Bacterial species are delineated by DNA-DNA hybridization (DDH) which has been the standard for prokaryotic classification. Classification and identification are conducted according to a scientific approach and validated by peer review. Progress in the technologies available for the investigation of genomic data during the last 20 years has had a tremendous impact on prokaryotic classification studies. The key trait of taxonomical entities which interest the plant health manager is the ability to cause diseases of concern for plant health in a designed pest risk area. It will be a bacterial species when the disease is caused by a taxon that corresponds to the species concept. In many cases, taxonomical entities that cause plant diseases do not match with species delineations. Numerous diseases are causes by infra specific entities and the pathovar concept was proposed by the ad hoc committee on plant pathogenic bacteria. The term pathovar is used to refer to a set of strains with the same or similar characteristics, differentiated at infrasubspecific level from other strains of the same species or subspecies on the basis of distinctive pathogenicity to one or more plant hosts (Young et al. 1991, 2001). Classification methods and bacterial species concepts have advanced and their application to plant pathogenic bacteria has resulted both in the elevation of pathovars to subspecies and in the transfer of pathovars from one species to another.

Here we describe the importance of correct identification of bacteria subject to regulation. The examples we used have been the subject of EFSA opinions in the last 2 years. The PLH Panel on Plant Health was requested to a perform a PRA on *Dickeya dianthicola* a regulated organism previously called *Erwinia chrysenthemi* pv. *dianthicola*, on carnation and *Xanthomonas citri* pv. *citri* and *X. citri pv. aurantifolii* regulated as *Xanthomonas campestris* (all strains pathogenic on citrus). The taxonomy of these bacterial groups has fundamentally changed since the establishment of quarantine list in the 2000/29/EC directive and an updating of taxonomical identity of the pest was requested.

Dickeya dianthicola

Erwinia chrysanthemi was first included in the genus *Erwinia* as a pathogen of chrysanthemum (Burkholder et al. 1953). Subsequently, it was found to infect a wide variety of plants (Samson et al. 2005). In 1984, the species was subdivided into six pathovars – *dianthicola, zeae, chrysanthemi, parthenii, paradisiaca* and *dieffenbachiae* – according to their host specificity (Lelliot and Dickey 1984). In 1998, *E. chrysanthemi* was moved to the genus *Pectobacterium* based on 16S analysis (Hauben et al. 1998). In 2005, using 16S, DNA–DNA hybridisation and biochemistry, *P. chrysanthemi* was moved into a new genus called *Dickeya* (Samson et al. 2005), which comprises six species, namely *dianthicola, dadantii, zeae, chrysanthemi, paradisiaca* and *dieffenbachiae*. These species fall largely in line with the different *E. chrysanthemi* pathovars and show a degree of host specificity. In 2011, the species *D. dieffenbachiae* was moved within the species *D. dadantii* (Brady et al. 2011). More recent analyses have revealed the presence of a number of potentially new *Dickeya* spp. (including '*D. solani*'), but to date none of these has been officially recognised (Parkinson et al. 2009; Slawiak et al. 2009). *Dickeya* spp. are soft rotting pathogens that cause disease primarily through the production of large quantities of various plant cell wall-degrading enzymes.

According to Samson et al. (2005), the description of *Dickeya dianthicola* follows' *Pectobacterium parthenii-dianthicola*, described on *Dianthus* sp. (Hellmers 1955). Strains belong to ex *Pectobacterium chrysanthemi* biovars 1, 7 and 9. According to several phylogenetic studies including 16S, *recA* and *dnaX* sequence analysis, DNA–DNA hybridisation and REP-PCR, *D. dianthicola* is most closely related to *D. dadantii* and exhibits little diversity between strains, with no obvious delineation between isolates from different host plants (Samson et al. 2005; Parkinson et al. 2009; Sławiak et al. 2009). Using DNA–DNA hybridisation, Samson et al. (2005) demonstrated that biovars 1, 7 and 9 exhibit 71–100 % relatedness to the *D. dianthicola* type strain (CFBP 2012), with other biovars showing only 23–53 % relatedness to this strain. In another study using *dnaX* sequence analysis and biochemical assays, Sławiak et al. (2009) showed that biovars 1 and 7 were associated with the *D. dianthicola* type strain.

D. dianthicola was first detected causing slow wilting and stunting on *Dianthus* (Hellmers 1958; Dickey 1979; Samson et al. 2005; Parkinson et al. 2009). Janse and Ruissen (1988) showed that a strain isolated from *Dianthus caryophyllus* in the UK was not able to cause disease in *Kalanchoe blossfeldiana*, suggesting a degree of specialisation between strains/biovars. The host range of *D. dianthicola* is quite large. Other hosts in which *D. dianthicola* has been shown to cause disease include potato, tomato, chicory, begonia, *Dahlia* spp. (including *Dahlia pinnata* and *Dahlia variabilis*), artichoke, *Kalanchoe* spp. (including *Kalanchoe blossfeldiana*) (Dickey 1979; Samson et al. 2005; Sławiak et al. 2009; Laurila et al. 2010), hyacinthus and sedum (Samson and Nassan-Agha 1978; Sławiak et al. 2009; van der Helm 2009). In dahlia, *Dickeya* has been isolated from infected tubers (van Doorn et al. 2008). Recently, in infection experiments it was shown that *D. dianthicola* can cause wilting of dahlia shoots and bursting of tubers ('ploffers'). When isolates were obtained from natural symptomatic dahlia plants, *D. dianthicola* was found to be present (van Leeuwen et al. 2012). More recently, there have been a number of reports concerning the isolation of *Dickeya* spp. from weeds (Toth et al. 2011).

Xanthomonas citri pv. *citri*

The Council Directive 2000/29/EC used the *Xanthomonas* nomenclature that was in place before the reclassification of the genus in 1995 (Dye and Lelliott 1974; Vauterin et al. 1995). Subsequent international research efforts were done later on *Xanthomonas* taxonomy (Vauterin and Swings 1997; Rademaker et al. 2000; Young et al. 2008; Rodriguez et al. 2012). The strains of *X. campestris* pathogenic to citrus have been reclassified as distinct species and also differ markedly in terms of symptomatology, host range and economical significance.

Before 1995, two taxonomical entities caused citrus bacterial canker disease. (i) *X. campestris* pv. *citri* pathotype A is the causal agent of Asiatic citrus canker. This pathogen groups into genetic cluster 9.5 of *X. axonopodis* sensu Vauterin et al. (1995) (Rademaker et al. 2000). It has been reclassified as *X. citri* pv. *citri* (synonyms *X. citri* subsp. *citri* or *X. axonopodis* pv. *citri*) (Ah-You et al. 2009; Schaad et al. 2006; Vauterin et al. 1995). Variants of *X. citri* pv. *citri*, which are phylogenetically very close but pathologically distinct in terms of host range, have been reported as pathotypes A*/Aw (Bui Thi Ngoc et al. 2009, 2010; Sun et al. 2004; Vernière et al. 1998). (ii) *X. campestris* pv. *citri* pathotype B/C has been reported as the causal agent of South American citrus canker. These strains group into genetic cluster 9.6 of *X. axonopodis* sensu Vauterin et al. (1995) and have been reclassified in 2006 as *X. fuscans* subsp. *aurantifolii* (synonyms *X. citri* pv. *aurantifolii* or *X. axonopodis* pv. *aurantifolii*) (Ah-You et al. 2009; Schaad et al. 2006; Vauterin et al. 1995). However, recent data did not support *X. fuscans* as a separate species (Young et al. 2008) and suggested that it may be a later heterotypic synonym of *X. citri* (Ah-You et al. 2009). This was further confirmed by a pangenomic phylogeny of the genus *Xanthomonas* (Rodriguez et al. 2012).

It is now recognised that two pathovars within *Xantomonas citri* (pv. *citri* and pv. *aurantifolii*) cause citrus bacterial canker and should be taken into consideration to perform a pest risk assessment on citrus bacterial canker-causing bacteria.

Additionally, two taxonomic entities cause a symptomatology markedly different disease from that of citrus canker. Symptoms consist of flat, water-soaked spots developing into necrotic lesions and are most often visible on citrumelo rootstock (*Citrus paradisi* x *Poncirus trifoliata*) and its parents (Graham and Gottwald 1991). *X. alfalfae* subsp. *citrumelonis* should therefore be considered a pathogen distinct from *X. citri* and the associated disease, citrus bacterial spot, a disease distinct from citrus canker. Similar to *X. alfalfae* subsp. *citrumelonis* in terms of symptomatology, *X. campestris* pv. *bilvae* produces flat, water-soaked spots developing into necrotic lesions on Aegle, Feronia and Mexican lime (*Citrus aurantifolia*) (Bui Thi Ngoc et al. 2010; Patel et al. 1953). A single report of this pathogen has been made from India (Patel et al. 1953), but not further confirmed. There are no indications of outbreaks caused by this bacterium worldwide. These strains have been reclassified in 2010 as *X. citri* pv. *bilvae* (Bui Thi Ngoc *et al.* 2010).

In the laboratory, all xanthomonads responsible for the above-listed bacterial diseases of Citrus can be readily distinguished on the basis of several molecular techniques such as rep-PCR (Egel et al. 1991; Rademaker et al. 2005), Amplified Fragment Length Polymorphism (AFLP) (Janssen et al. 1996; Bui Thi Ngoc et al. 2010) and MultiLocus Sequence Analysis (MLSA) (Almeida et al. 2010; Bui Thi Ngoc et al. 2010; Young et al. 2008).

Viruses and Viroids

The ability to detect and/or identify plant viruses and viroids has much benefited, as in all other plant pathology fields, from advances in molecular diagnostic methods. In particular, the characterization of the agents responsible for listed viral diseases of unknown etiology, which could previously only be detected through biological indexing on indicator plants, now permits their easy identification using molecular tools. However, a few examples of quarantine diseases of suspected viral origin still remain in the EU plant health legislation, such as the "vein enation woody gall" and the "blight and blight-like" diseases, that are banned on citrus plants in the annexes of the EU plant health Directive 2000/29/EC. For the first of these, a new *Enamovirus* has recently been tentatively associated with the disease (Vives et al. 2013), which may in time simplify diagnostics and legislation compliance.

As for other agents, evolution of the taxonomy, with the discovery of new species or the splitting of old ones into separate taxa may complicate not only the interpretation of the legislation but also risk analysis, because it is often difficult to evaluate which currently recognized species is addressed in old literature records. Such a situation was recently encountered during an EFSA assessment of risks posed by several members of the *Secoviridae* family. Indeed, for one of these agents, *Tomato black ring virus* (TBRV, genus *Nepovirus*), the previously recognized E (English) and S (Scottish) serotypes are now considered as belonging to two distinct species, the former as *Tomato black ring virus* and the later as *Beet ringspot virus*, a name reintroduced for the occasion (Sanfaçon et al. 2012). However, in this particular situation, the fact that the biological properties impacting risk assessment of these two viruses are largely similar allowed them to be assessed together, limiting both effort and confusion.

A similarly complex situation was encountered during a recent mandate in which EFSA developed a risk assessment addressing the whole *Tospovirus* genus. Analysis of the literature indicated that only eight species were fully validated by the International Committee for the Taxonomy of viruses (ICTV), while 15 species were known but not yet internationally approved and a further three species were of so recent discovery that they were not mentioned in the most recent ICTV report available (Plyusnin et al. 2012; EFSA Panel on Plant Health 2012a). Two other species identified in the literature were in fact mere synonyms of valid species. In addition to this already complex situation, the existence of serological cross-reactions and of recombinants between *Tospovirus* genus members (Webster et al. 2011) has led in the past to misidentifications, complicating the interpretation of old publications (EFSA Panel on Plant Health 2012a).

Another situation in which taxonomy and identification issues can be encountered when assessing the risks posed by viruses concerns the taxonomic identification of vector species. Considerable risks can be associated with the introduction of exotic vectors of plant viruses: newly introduced vectors can facilitate the spread of exotic viruses or of endemic ones. Likewise, the pre-existence in an area of efficient vectors may greatly increase the risks of establishment and spread of exotic viruses upon entry. The recent pest categorization of tospoviruses (EFSA Panel on Plant Health 2012a) encountered difficulties in two areas: problems of misidentification (and potentially of misreporting) of thrips vector species, as for example in cases of confusion between *Thrips palmi* and *Thrips flavus* (Mound 1996), and problems of proper experimental validation, using clearly identified viral isolates and thrips populations, of the vector status of some thrips species (Whitfield et al. 2005).

As indicated in the introduction, an interesting situation concerns cases in which a whole group of related or unrelated organisms are listed together on the basis of a common biological property felt important for risk assessment. This concerns, for example, the listing in Directive 2000/29/EC of viruses transmitted by *Bemisia tabaci*, which was met during a recent assessment by EFSA of the risks posed by *Bemisia tabaci* (EFSA Panel on Plant Health 2013c). The situation is particularly complex in this instance because these viruses form a very heterogenous ensemble, containing both circulatively transmitted agents (*Begomovirus* genus), which remain associated with the insects for extended periods of time and for which viruliferous insects could also be considered as a potential virus introduction pathway, and non-circulative agents (genera *Crinivirus*, *Ipomovirus, Carlavirus* and *Torradovirus*) for which the risk of introduction through movement of vectors is much more limited. A further complexity layer is added by the variability in vector efficiency observed for some of these viruses between members of the *Bemisia tabaci* species complex (McGrath and Harrison 1995).

In some cases, the risk analysis can however benefit from the collective analysis of a taxonomic group of viruses or viroids. In a recent assessment of the risk of pospiviroids for solanaceous crops, the EFSA PLH Panel concluded that, owing to their largely similar biological and epidemiological features, the risk and the risk mitigation measures could be analyzed collectively for nine *Pospiviroid* species (*Potato spindle tuber viroid*, *Citrus exocortis viroid*, *Columnea latent viroid*, *Mexican papita viroid*, *Tomato apical stunt viroid*, *Tomato chlorotic dwarf viroid*, *Tomato planta macho viroid*, *Chrysanthemum stunt viroid* and *Pepper chat fruit viroid*) affecting solanaceous crops (EFSA Panel on Plant Health 2011a and references therein). These species were seen as sharing the following key characteristics for risk assessment: (1) availability of reliable molecular detection and identification/discrimination methods at species and at genus level (EFSA Panel on Plant Health 2011a; Torchetti et al. 2012); (2) although *Potato spindle tuber viroid* is the only pospiviroid infecting potato in nature, all eight other species can be transmitted to potato under experimental conditions and cause disease symptoms; (3) natural infection in tomato is recorded for seven species and, under experimental conditions, for the remaining two; (4) several of these viroids are carried asymptomatically by several ornamental species, in particular vegetatively propagated solanaceous ones; (5) seed transmission is demonstrated in tomato for four species, and in pepper for one, with a potential contribution to outbreaks in solanaceous vegetable crops; (6) several pospiviroids can be transmitted within short distances by pollen, bumblebees and, for PSTVd in specific mixed-infection conditions, by aphids; (7) the most effective transmission mechanism within a crop remains mechanical transmission during crop handling by workers. In the absence of precise experimental data for each of these properties for all of the nine pospiviroids, taxonomic closeness and the fact that the properties appeared shared by all or most pospiviroids in which they had been investigated allowed an assessment "by proxy" (but with higher uncertainty levels) that could not have been achieved if relying only on direct experimental data.

Nematodes

Plant-parasitic nematodes are of great economic importance. When left unchecked and uncontrolled, they may have an adverse impact on plant health and can cause severe yield losses. It has been estimated in monetary terms that about $US80 billion is lost per year as a result of plant nematode damage (Nicol et al. 2011; Jones et al. 2013). Detrimental effects of plant parasitic nematodes are commonly underestimated because the symptoms vary with environmental conditions and the plant growth stage and because growers, agronomists and pest management specialists are usually not aware of the damage caused by nematodes.

With regard to their devastating impact on host plants, the accurate and timely identification of plant parasitic nematodes is indispensable and has important implications for many areas, including systematics (taxonomy and phylogeny), population genetics, ecology and epidemiology and is a prerequisite for creating effective management strategies (Gasser 2001).

Potato Cyst Nematode

Potato cyst nematodes (PCN), *Globodera rostochiensis* and *G. pallida* are the most important nematode threat to potato production worldwide. If left uncontrolled, these nematodes are capable of causing a 75 % reduction in potato yields (Seinhorst 1982). They are listed in the EU Plant Health Directive 2000/29/EC and are also subject of the EU Council Directive 2007/33/EC. PCN originated from South America, from where they were introduced to Europe in the middle of the nineteenth century (Evans et al. 1975) in soil adhering to potato seed tubers (Hockland et al. 2012). From Europe, PCN were then spread into most other potato growing areas around the world.

Diagnostics of *Globodera* spp. is based on morphological characteristics of cysts and juveniles. However, this type of diagnostics can be time-consuming and generally requires skilled personnel with specialized taxonomic expertise. Furthermore, visual identification is not always unambiguous because some of diagnostic characteristics may overlap between various populations and different species (Fleming and Powers 1998; EFSA Panel on Plant Health 2012b). Therefore, many different biochemical techniques have been developed to separate *Globodera rostochiensis* and *G. pallida*, or to differentiate a wide range of related species. In addition, several authors investigated differences between European and South American populations of PCN. However no molecular method to distinguish these populations or to identify populations with specific virulence characteristics has been developed so far (EFSA Panel on Plant Health 2012b).

Based on biological and molecular studies it is evident that PCN present in Europe represent a minor subset of full biological diversity present in South America (Franco and Evans 1978; Plantard et al. 2008; Grenier et al. 2010) and that the range of virulence present in South America is far greater than that present in European PCN populations (EFSA Panel on Plant Health 2012b).

The most important and effective option to control PCN is the use of resistant potato cultivars which have been bred against PCN populations currently present in Europe (EFSA Panel on Plant Health 2012b). Almost all potato cultivars that are grown in Europe and are resistant to *G. rostochiensis* and *G. pallida* derive this resistance from the H1 and H3 gene, respectively. Unfortunately, a range of South American populations of both PCN species are able to overcome the H1, H2 and H3 resistance genes. Each of South American populations tested overcomes at least one of these resistance sources (Franco and Evans 1978; EFSA Panel on Plant Health 2012b). It is therefore evident that the introduction of new virulent PCN strains from South America into Europe poses a potential risk for European potato production (Hockland et al. 2012).

Xiphinema Group

Despite their economic importance some important nematodes, including species of *Xiphinema americanum* group, are not included on the list of scientifically and economically most important plant parasitic nematodes that was presented by Jones et al. (2013).

The *Xiphinema americanum* species complex attracts special attention because of the taxonomic confusion and because it presents a very important source of virus vectoring. Six species of this group (*X. americanum* sensu stricto, *X. californicum*, *X. bricolense*, X. *intermedium*, *X. rivesi*, *X. tarjanense*) are known to transmit economically and phytosanitary very important nepoviruses that are included in the EU and EPPO lists of quarantine organisms. In North America, the four nepoviruses Cherry rasp leaf (CRLV), Peach rosette mosaic (PRMV), Tobacco ringspot (TRSV) and Tomato ringspot (ToRSV) that cause substantial damage to a wide range of crops are known to be transmitted by nematode species belonging to the *Xiphinema americanum* group (Decraemer and Robbins 2007). Due to their importance and in order to prevent the introduction of non-indigenous plant viruses into Europe, non-European populations of *Xiphinema americanum* sensu lato are listed in Annex IAI of EC Council Directive 2000/29/EC.

Many species from the *Xiphinema americanum* group has worldwide distribution and some of them are also present in Europe. However none of the European populations except the Slovenian population of *X. rivesi* has been demonstrated to transmit the quarantine-listed nepoviruses (EPPO 2009; Širca et al. 2007). The ability of Slovenian population of *X. rivesi* to transmit TRSV and ToRSV to bait plants has been recently confirmed (Širca et al. 2007).

Species identification within the *X. americanum* group is of particular importance for phytosanitary regulation; however, differentiation of species within this group remains problematic despite some attempts to create useful and definitive morphological and molecular tools for species detection and diagnosis. Many characters used for species identification are variable because of the influence of environmental factors on growth and because geographical isolation of widespread species may promote genetic drift (Taylor and Brown 1997). Consequently, characters that are essential for unambiguous identification often overlap, making diagnostics difficult or even prevented. The total number of valid species of *X. americanum* group is therefore still debated (Luc et al. 1998; Lamberti et al. 2000). With respect to the group of specialist who deal with taxonomy of *X. americanum* group 34 (Luc et al. 1998), 38 (Coomans et al. 2001) or 49 (Lamberti et al. 2004) putative species are included in the *Xiphinema americanum*-group.

The whole story related to the taxonomic controversy of the *X. americanum* group dates back to seventies when Lima (1965) and Tarjan (1969) suggested that populations identified as *X. americanum* Cobb actually represented a species complex. After studying thousands of specimens and several hundred of populations classified by various authors within the *X. americanum* group, Lamberti and Bleve-Zacheo (1979) restricted the definition of *X. americanum* sensu stricto and divided the species group into six groups containing 15 new species among a total of 25 species that could be recognized as belonging to the species complex. In 1991, Lamberti and Carone published dichotomous identification key for 38 species of *X. americanum* group. Since then, the taxonomy of this group has been intensively studied and debated by many authors (Luc et al. 1998; Lamberti et al. 2000, 2004; Coomans et al. 2001; Luc and Baujard 2001; He et al. 2005; Gutiérrez-Gutiérrez et al. 2012), but still remains controversial.

Due to very small differences between several described species of *X. americanum* group and little information on intraspecific variability, distinguishing the members of this complicated group remains problematic. However, availability of molecular techniques may help to provide tools for differentiating *Xiphinema americanum* group species and can significantly improve and facilitate the routine identification of these nematodes (Gutiérrez-Gutiérrez et al. 2012). Polyphasic identification combining molecular techniques with morphology and measurements for species diagnosis is therefore essential for precise and unequivocal species identification within *X. americanum* group (Gutiérrez-Gutiérrez et al. 2011, 2012). Continual improvement of the accuracy, speed, sensitivity and quality of sample analysis will increase feasibility of designing molecular diagnostics to distinguish plant parasitic nematodes (Blok 2005).

However, there are still practical constraints concerning sampling which limit practical use of diagnostic procedures. Sampling for the presence of plant parasitic nematodes that are mostly soil-dwelling pathogens is difficult and, when sampling a field, only a limited amount of soil can be processed for nematode detection.

Conclusion

The case studies used in this Chapter provide ample evidence for the problems posed for risk assessment when taxonomic and nomenclature changes affect the organisms listed in the Council Directive of 2009. These are clearly shown in the risk assessments carried out for the two fungal pathogens *Stagonosporopsis chrysanthemi* and *Dothistroma septosporum* where neither of these named organisms corresponded to the regulated organisms. For the latter organism there is no regulation in place for the related species *D. pini*, and which may have been mistakenly diagnosed in the past. A similar situation arises with the bacterium *Dickeya dianthicola*. For *Xanthomonas citri* pv. *citri*, there has been almost continuous confusion and revised nomenclature since 2000, when the regulation put in place at that time did not even correspond to the then-accepted nomenclature. For viruses and viroids, issues relating to taxonomy and nomenclature are even more complex, with the almost complete reliance on molecular methods for diagnosis, perhaps different species concepts to other taxa, and the added complication of vector transmission. For this reason, risk assessments have mostly been approached based on taxonomic groupings as illustrated by the tospoviruses and the pospiviroids. With vector transmission there is the further complication that there has been an equally-changing taxonomy and nomenclature, as illustrated by the *Bemisia* whitefly and *Xiphinema* nematode examples. The importance of subspecific variation in risk assessment is illustrated by potato cyst nematode where the issue of concern was the different virulence characteristics of South American vs European populations. Improved diagnostic techniques can also simply the risk assessments as in the case of *P. fragariae*, through its elevation to specific status as the only *Phytophthora* causing red core disease of strawberry. There are additional problems when organisms previously unknown to science arise as plant health concerns, typified in particular by *Phytophthora ramorum*, not named or listed in 2000 but which subsequently was subject to EU emergency measures. In this case the improved diagnostic procedures that have been developed in response have been invaluable not only in improved and rapid identification, but also in tracking the populations once introduced and the likely pathways involved in spread.

Acknowledgments Authors thank the following working group members, hearing experts and EFSA staff involved in the scientific opinions mentioned in this Chapter: R. Baker, T. Been, P. Bielza, P. Bonants, C. Brasier, A. Brown, P.J. De Barro, H. De Gruyter, F. Di Serio,

E. Dormannsne Simon, J. Duncan, R. Flores, A. Fox, M. Garbelotto, N. Gauthier, G. Gilioli, T. Goumperis, J. Gregoire, E. Grenier, I, Holb, S. Hockland, G. Hollo, M. Ioannou, D. Janssen, J. Jones, O.E. Karadjova, V. Kertesz, S. Kozelska, O. Mosbach-Schulz, M. Navajas, S. Pasquali, M. Pautasso, D. Peters, A. Porta Puglia, T. Rafoss, A. Rortais, M. Ruffini, J. Schans, R. Singh, A. Sletten, S. Tramontini, A. Urbaneja, E. van de Weg, J.C. Van Lenteren, J.T.J. Verhoeven, I. Vloutoglou, S. Vos, S. Winter

The present paper is published under the sole responsibility of the authors and may not be considered as a scientific output by the European Food Safety Authority (EFSA). The position and opinions presented in this publication are those of the authors alone and do not necessarily represent the views of EFSA.

References

Ah-You N, Gagnevin L, Grimont PAD, Brisse S, Nesme X, Chiroleu F, Bui Thi Ngoc L, Jouen E, Lefeuvre P, Vernière C, Pruvost O (2009) Polyphasic characterization of *Xanthomonas* pathogenic to *Anacardiaceae* and their relatedness to different *Xanthomonas* species. Int J Syst Evol 59(2):306–318

Almeida NF, Yan S, Cai R, Clarke CR, Morris CE, Schaad NW, Schuenzel EL, Lacy GH, Sun X, Jones JB, Castillo JA, Bull CT, Leman S, Guttman DS, Setubal JC, Vinatzer BA (2010) PAMDB, a multilocus sequence typing and analysis database and website for plant-associated microbes. Phytopathology 100:208–215

Aveskamp MM, de Gruyter J, Crous PW (2008) Biology and recent developments in the systematics of *Phoma*, a complex genus of major quarantine significance. Fungal Divers 31:1–18

Aveskamp MM, Verkley GJM, De Gruyter J, Murace MA, Perelló A, Woudenberg JH, Groenwald JZ, Crous PW (2009) DNA phylogeny reveals polyphyly of *Phoma* section *Peyronellaea* and multiple taxonomic novelties. Mycologia 101(3):363–382

Aveskamp MM, De Gruyter J, Woudenberg JHC, Verkley GJM, Crous PW (2010) Highlights of the Didymellaceae: a polyphasic approach to characterise *Phoma* and related *pleosporalean* genera. Stud Mycol 65(1):1–60

Baker KF, Dimock A, Davis LH (1949) Life history and control of the ascochyta ray blight of *chrysanthemum*. Phytopathology 39:789–805

Barnes I, Crous PW, Wingfield BD, Wingfield MJ (2004) Multigene phylogenies reveal that red band needle blight of *Pinus* is caused by two distinct species of *Dothistroma, D. septosporum* and *D. pini*. Stud Mycol 50:551–565

Barnes I, Kirisits T, Akulov A, Chhetri DB, Wingfield BD, Bulgakov TS, Wingfield MJ (2008) New host and country records of the *Dothistroma* needle blight pathogens from Europe and Asia. For Pathol 38:178–195

Barnes I, Kirisits T, Wingfield MJ, Wingfield BD (2011) Needle blight of pine caused by two species of *Dothistroma* in Hungary. For Pathol 41:361–369

Blok VC (2005) Achievements in and future prospects for molecular diagnostics of plant-parasitic nematodes. Can J Plant Pathol 27:176–185

Bonants PMJ, Van Gent-Pelzer MPE, Hooftman R, Cooke DEL, Guy DC, Duncan JM (2004) A combination of baiting and different PCR formats, including measurement of real-time quantitative fluorescence, for detection of *Phytophthora fragariae* in strawberry plants. Eur J Plant Pathol 110:689–702

Boutet X, Vercauteren A, Heungens K, Laurent F, Chandelier A (2010) Oospores progenies from *Phytophthora ramorum*. Fungal Biol 114:369–378

Brady C, Cleenwerck I, Denman S, Venter S, Rodríguez-Palenzuela P, Couthino TA, Vos P (2011) Proposal to reclassify. *Brenneria quercina*, (Hildebrand & Schroth 1967) Hauben et al. 1999 into a novel genus, *Lonsdalea* gen. nov., as *Lonsdalea quercina* comb. nov., descriptions of

Lonsdalea quercina subsp. *quercina* comb. nov., *Lonsdalea quercina* subsp. *iberica* subsp. nov. and *Lonsdalea quercina* subsp. *britannica* subsp. nov., emendation of the description of the genus Brenneria, reclassification of Dickeya dieffenbachiaeas *Dickeya dadantii* subsp. *dieffenbachiae* comb. nov., and emendation of the description of *Dickeya dadantii*. Int J Syst Evol Microbiol 62:1592–1602

Brasier CM (2003) Sudden oak death exhibits transatlantic differences. Mycol Res 107:258–259

Brasier CM, Kirk S (2004) Production of gametangia by *Phytophthora ramorum* in vitro. Mycol Res 108:823–827

Brasier CM, Vettraino AM, Chang TT, Vannini A (2010) *Phytophthora lateralis* discovered in an old growth Chamaecyparis forest in Taiwan. Plant Pathol 59:595–603

Bui Thi Ngoc L, Vernière C, Jarne P, Brisse S, Guérin F, Boutry S, Gagnevin L, Pruvost O (2009) From local surveys to global surveillance: three high throughput genotyping methods for the epidemiological monitoring of *Xanthomonas citri* pv. *Citri pathotypes*. Appl Environ Microbiol 75:1173–1184

Bui Thi Ngoc L, Vernière C, Jouen E, Ah-You N, Lefeuvre P, Chiroleu F, Gagnevin L, Pruvost O (2010) Amplified fragment length polymorphism and multilocus sequence analysis-based genotypic relatedness among pathogenic variants of *Xanthomonas citri* pv. citri and *Xanthomonas campestris* pv. bilvae. Int J Syst Evol Microbiol 60:515–525

Burkholder WH, McFadden LA, Dimock AW (1953) A bacterial blight of chrysanthemums. Phytopathology 43:522–526

Cooke DEL, Drenth A, Duncan JM, Wagels G, Brasier CM (2000) A molecular phylogeny of *Phytophthora* and related Oomycetes. Fungal Genet Biol 30:17–32

Coomans A, Huys R, Heyns J, Luc M (2001) Character analysis, phylogeny, and biogeography of the genus *Xiphinema* Cobb, 1913 (Nematoda: Longidoridae). Annales Musée Royal de L'Afrique Centrale (Zoologie) 287:1–289

De Gruyter JD, Boerema GH, Aa HA (2002) Contributions towards a monograph of *Phoma* (Coelomycetes). VI-2. Section Phyllostictoides: outline of its taxa. Persoonia 18(1):1–53

De Gruyter J, Aveskamp MM, Woudenberg JHC, Verkley GJM, Groenewald JZ, Crous PW (2009) Molecular phylogeny of *Phoma* and allied anamorph genera: towards a reclassification of the *Phoma* complex. Mycol Res 113:508–519

De Gruyter J, Woudenberg JHC, Aveskamp MM, Verkley GJM, Groenewald JZ, Crous PW (2010) Systematic reappraisal of species in *Phoma* section *Paraphoma, Pyrenochaeta* and *Pleurophoma*. Mycologia 102:1066–1081

De Gruyter J, Woudenberg JHC, Aveskamp MM, Verkley GJM, Groenewald JZ, Crous PW (2013) Redisposition of *Phoma*-like anamorphs in *Pleosporales*. Stud Mycol 75:1–36

Decraemer W, Robbins RT (2007) The who, what and where of Longidoridae and Trichodoridae. J Nematol 39:295–297

Dickey RS (1979) *Erwinia chrysanthemi*: a comparative study of phenotypic properties of strains from several hosts and other *Erwinia* species. Phytopathology 69:324–329

DOE JGI (DOE Joint Genome Institute) (2012) Dothideomycete sequencing and comparative analysis. US Department of Energy, Office of Science. University of California, USA. Available online at: http://genome.jgi.doe.gov/programs/fungi/fungal-projects.jsf

Duncan JM (2001) Redcore. Third and final year progress report to EU DGXII (SMT4-CT97-2162) 1–42

Duncan JM, Kennedy DM, Scott PH (1991) Relationships between non-papillate, soilborne species of *Phytophthora* I: root rot of raspberry. In Phytophthora eds Publi. Cambridge, University Press for British Mycological Society, pp 139–147

Dye DW, Lelliott RA (1974) Genus II. *Xanthomonas* Dowson 1939, 187. In: Bergey's manual of determinative bacteriology. Williams and Wilkins, Baltimore, pp 243–249

EFSA Panel on Plant Health (2011a) Scientific opinion on the assessment of the risk of solanaceous pospiviroids for the EU territory and the identification and evaluation of risk management options. EFSA J 9:133 (2330)

EFSA Panel on Plant Health (2011b) Scientific opinion on the pest risk analysis on *Phytophthora ramorum* prepared by the FP6 project RAPRA. EFSA J 9:107 (2186)

EFSA Panel on Plant Health (2012a) Scientific opinion on the pest categorisation of the tospoviruses. EFSA J 10:101 (2772)

EFSA Panel on Plant Health (2012b) Scientific opinion on the risks to plant health posed by European versus non-European populations of the potato cyst nematodes *Globodera pallida* and *Globodera rostochiensis*. EFSA J 10:71 (2644)

EFSA Panel on Plant Health (2013a) Scientific opinion on the risks to plant health posed by *Stagonosporopsis chrysanthemi* (Stevens) Crous, Vaghefi and Taylor [*Didymella ligulicola* (Baker, Dimock and Davis) Arx var. *ligulicola*; syn. *Didymella ligulicola* (Baker, Dimock and Davis) Arx] in the EU territory, with identification and evaluation of risk reduction options. EFSA J 11:72 (3376)

EFSA Panel on Plant Health (2013b) Scientific opinion on the risk to plant health posed by *Dothistroma septosporum* (Dorog.) M. Morelet (*Mycosphaerella pini* E. Rostrup, syn. *Scirrhia pini*) and *Dothistroma pini* Hulbary to the EU territory with the identification and evaluation of risk reduction options. EFSA J 11:173 (3026)

EFSA Panel on Plant Health (2013c) Scientific opinion on the risks to plant health posed by *Bemisia tabaci* species complex and viruses it transmits for the EU territory. EFSA J 11:302 (3162)

EFSA Panel on Plant Health (2014) Scientific opinion on the risks to plant health posed by *Phytophthora fragariae* Hickman var. *fragariae* in the EU territory, with the identification and evaluation of risk reduction options. EFSA J 12(1):63 (3539)

Egel DS, Graham JH, Stall RE (1991) Genomic relatedness of *Xanthomonas campestris* strains causing diseases of citrus. Appl Environ Microbiol 57:2724–2730

Elliott M, Sumampong G, Varga A, Shamoun SF, James D, Masri S, Briere SC, Grünwald NJ (2009) PCR-RFLP markers identify three lineages of the North American and European populations of *Phytophthora ramorum*. For Pathol 39:266–278

EPPO (2009) *Xiphinema americanum* sensu lato. Bulletin OEPP/EPPO Bulletin 39:382–392

Evans K, Franco J, De Scurrah MM (1975) Distribution of species of potato cyst nematodes in South America. Nematologica 21:365–369

Fleming CC, Powers TO (1998) Potato cyst nematodes: species, pathotypes and virulence concepts. In: Marks RJ, Brodie BB (eds) Potato cyst nematodes: biology, distribution and control. CABI Publishing, Wallingford, pp 51–57

Franco J, Evans K (1978) Multiplication of some South American and European populations of potato cyst Nematodes on potatoes possessing the resistance genes H_1, H_2 and H_3. Plant Pathol 27:1–6

Garbelotto M, Rizzo DM (2005) A California-based chronological review (1995–2004) of research on *Phytophthora ramorum*, the causal agent of sudden oak death. Phytopathol Mediterr 44(2):127–143

Gasser RB (2001) Identification of parasitic Nematodes and study of genetic variability using PCR approaches. In: Kennedy MW, Harnett W (eds) Parasitic nematodes, molecular biology, biochemistry and Immunology. CABI Pub, New York, pp 53–82

Goss EM, Carbone I, Gruenwald NJ (2009) Ancient isolation and independent evolution of the three clonal lineages of the exotic sudden oak death pathogen *Phytophthora ramorum*. Mol Ecol 18:1161–1174

Graham JH, Gottwald TR (1991) Research perspectives on eradication of citrus bacterial diseases in Florida. Plant Dis 75:1193–1200

Grenier E, Fournet S, Petit E, Anthoine G (2010) A cyst nematode "species factory" called the Andes. Nematology 12:163–169

Groenewald M, Barnes I, Bradshaw RE, Brown AV, Dale A, Groenewald JZ, Lewis KJ, Wingfield BD, Wingfield MJ, Crous PW (2007) Characterization and distribution of mating type genes in the *Dothistroma* needle blight pathogens. Phytopathology 97:825–834

Gutiérrez-Gutiérrez C, Castillo P, Cantalapiedra-Navarrete C, Landa BB, Derycke S, Palomares-Rius JE (2011) Genetic structure of *Xiphinema pachtaicum* and X. index populations based on mitochondrial DNA variation. Phytopathology 101:1168–1175

Gutiérrez-Gutiérrez C, Cantalapiedra-Navarrete C, Decraemer W, Vovlas N, Prior T, Palomares Rius JE, Castillo P (2012) Phylogeny, diversity, and species delimitation in some species of the *Xiphinema americanum*-group complex (Nematoda: Longidoridae), as inferred from nuclear and mitochondrial DNA sequences and morphology. Eur J Plant Pathol 134:561–597

Hansen EM (2010) Some challenges of recognizing invasive *Phytophthoras* and finding their origins. In: Proceedings of the 4th Sudden Oak Death Symposium, USDA Forest Service, Albany, CA, USA, pp 17–21

Hauben L, Moore ERB, Vauterin L, Steenackers M, Mergaert J, Verdonck L, Swings J (1998) Phylogenetic position of phytopathogens within the Enterobacteriaceae. Syst Appl Microbiol 21:384–397

He Y, Subbotin S, Rubtsova TV, Lamberti F, Brown DJF, Moens M (2005) A molecular phylogenetic approach to Longidoridae (Nematoda: Dorylaimida). Nematology 7:111–124

Hellmers E (1955) Bacterial wilt of carnations. Gardeners'. Chronicle 137(3567):194–195

Hellmers E (1958) Four wilt diseases of perpetual-flowering carnations in Denmark. Dansk Botanisk Arkiv 18:1–200

Hickman C (1940) The red core root disease of the strawberry caused by *Phytophthora fragariae* n. sp. J Pomol 18:89–118

Hockland S, Niere B, Grenier E, Blok V, Phillips M, Den Nijs L, Anthoine G, Pickup G, Viaene N (2012) An evaluation of the implications of virulence in non-European populations of *Globodera pallida* and *G. rostochiensis* for potato cultivation in Europe. Nematology 14(1):1–13

Ioos R, Fabre B, Saurat C, Fourrier C, Frey P, Marçais B (2010) Development, comparison, and validation of real-time and conventional PCR tools for the detection of the fungal pathogens causing brown spot and red band needle blights of pine. Phytopathology 100:105–114

Ivors KL, Hayden K, Bonants PJM, Rizzo DM, Garbelotto M (2004) AFLP and phylogenetic analyses of North American and European populations of *Phytophthora ramorum*. Mycol Res 108:378–392

Ivors KL, Garbelotto M, Vries IDE, Ruyter-Spira C, Hekkert BT, Rosenzweig N, Bonants P (2006) Microsatellite markers identify three lineages of *Phytophthora ramorum* in US nurseries, yet single lineages in US forest and European nursery populations. Mol Ecol 15:1493–1505

Janse JD, Ruissen MA (1988) Characterisation and classification of *Erwinia chrysanthemi* strains from several hosts in the Netherlands. Phytopathology 78:800–808

Janssen P, Coopman R, Huys G, Swings J, Bleeker M, Vos P, Zabeau M, Kersters K (1996) Evaluation of the DNA fingerprinting method AFLP as a new tool in bacterial taxonomy. Microbiology 142:1881–1893

Jeger M, Schans J, Lövei GL, Van Lenteren J, Navajas M, Makowski D, Stancanelli G, Tramontini S, Ceglarska EB (2012) Special issue: risk assessment in support of plant health. EFSA J 10(10):8 (s1012)

Jones JT, Haegeman A, Danchin EGJ, Gaur HS, Helder J, Jones MGK, Kikuchi T, Manzanilla-López R, Palomares-Rius JE, Wesemael WML, Perry RN (2013) Top 10 plant-parasitic nematodes in molecular plant pathology. Mol Plant Pathol 14:946–961

Kliejunas JT (2010) Sudden oak death and *Phytophthora ramorum*: a summary of the literature. Gen Tech Rep PSW-GTR-234. Albany, CA: U.S. Department of Agriculture, Forest Service, Pacific Southwest Research Station

Lamberti F, Bleve-Zacheo T (1979) Studies on *Xiphinema americanum sensu lato* with descriptions of fifteen new species (Nematoda, Longidoridae). Nematol Mediterr 7:51–106

Lamberti F, Carone M (1991) A dichotomous key for the identification of species of *Xiphinema* (Nematoda: Dorylaimida) within the *X. americanum* group. Nematol Mediterr 19:341–348

Lamberti F, Molinari S, Moens M, Brown DJF (2000) The *Xiphinema americanum*-group. I. Putative species, their geographical occurrence and distribution, and regional polytomous identification keys for the group. Russian J Nematol 8:65–84

Lamberti F, Hockland S, Agostinelli A, Moens M, Brown DJF (2004) The *Xiphinema americanum* group. III. Keys to species identification. Nematol Mediterr 32:53–56

Laurila J, Hannukkala A, Nykyri J, Pasanen M, Helias V, Garlant L, Pirhonen M (2010) Symptoms and yield reduction caused by *Dickeya* spp. strains isolated from potato and river water in Finland. Eur J Plant Pathol 126:249–262

Lelliot RA, Dickey RS (1984) Genus VII. Erwinia Winslow, Broadhurst, Buchanan, Krumwiede, Rogers and Smith 1920, Bergey's manual of systematic bacteriology. The Williams & Wilkins 1, Baltimore, p 469–476

Lima MB (1965) Studies on species of the genus *Xiphinema* and other nematodes. Ph.D. Thesis, University of London, pp 165

Luc M, Baujard P (2001) On specific determination within the *Xiphinema americanum* group (Nematoda: Longidoridae). Nematology 3:727–728

Luc M, Coomans A, Loof PAA, Baujard P (1998) The *Xiphinema americanum* group (Nematoda: Longidoridae). Observations on *Xiphinema brevicollum* Lordello & Da Costa, 1961 and comments on the group. Fundam Appl Nematol 21:475–490

Man in't Veld WA (2007) Gene flow analysis demonstrates that *Phytophthora fragariae* var. rubi constitutes a distinct species, *Phytophthora rubi* comb.nov. Mycopathologia 99:222–226

Mascheretti S, Croucher P, Vettraino A, Prospero S, Garbelotto M (2008) Reconstruction of the sudden oak death epidemic in California through microsatellite analysis of the pathogen *Phytophthora ramorum*. Mol Ecol 17:2755–2768

Mascheretti S, Croucher P, Kozanitas M, Baker L, Garbelotto M (2009) Genetic epidemiology of the sudden oak death pathogen *Phytophthora ramorum* in California. Mol Ecol 18:4577–4590

McGrath PF, Harrison BD (1995) Transmission of tomato leaf curl geminiviruses by *Bemisia tabaci*: effects of virus isolate and vector biotype. Ann Appl Biol 126:307–316

Mound LA (1996) The *Thysanoptera* vector species of tospoviruses. Acta Horticulturae 431:298–306

Newton AC, Duncan JM, Augustin NH, Guy DC, Cooke DEL (2010) Survival, distribution and genetic variability of inoculum of the strawberry red core pathogen, *Phytophthora fragariae* var. *fragariae*, in soil. Plant Pathol 59:472–479

Nicol JM, Turner SJ, Coyne DL, Den Nijs L, Hockland S, Maafi ZT (2011) Current nematode threats to world agriculture. In: Jones JT, Gheysen G, Fenoll C (eds) Genomics and molecular genetics of plant–nematode interactions. Springer, Heidelberg, pp 21–44

Parkinson NM, Stead D, Bew J, Heeney J, Tsor (Lahkim) L, Elphinstone J (2009) *Dickeya* species relatedness and clade structure determined by comparison of *recA* sequences. Int J Syst Evol Microbiol 59:2388–2393

Patel MK, Allayyanavaramath SB, Kulkarni YS (1953) Bacterial shot-hole and fruit canker of *Aegle marmelos Correa*. Curr Sci 22:216–217

Pethybridge SJ, Scott JB, Hay FS (2004) Genetic relationships among isolates of *Phoma ligulicola* from *pyrethrum* and *chrysanthemum* based on ITS sequences and its detection by PCR. Aust Plant Pathol 33:173–181

Pethybridge SJ, Hay FS, Clarkson RA, Groom T, Wilson CR (2008) Host range of Australian *Phoma ligulicola* var. *inoxydablis* isolates from *Pyrethrum*. J Phytopathol 156:506–508

Plantard O, Picard D, Valette S, Scurrah M, Grenier E, Mugniery D (2008) Origin and genetic diversity of Western European populations of the potato cyst nematode (*Globodera pallida*) inferred from mitochondrial sequences and microsatellite loci. Mol Ecol 17:2208–2218

Plyusnin AMQ, Beaty BJ, Elliott RM, Goldbach R, Kormelink R, Lundkvist Å, Schmaljohn CS, Tesh RB (2012) Bunyaviridae. In: King MJA, Adams MJ, Carstens EB, Lefkowitz EJ (eds) Virus taxonomy, ninth report of the international committee on taxonomy of viruses. Elsevier Academic Press, London, pp 725–741

Prospero S, Hansen EM, Grünwald NJ, Winton LM (2007) Population dynamics of the sudden oak death pathogen *Phytophthora ramorum* in Oregon from 2001 to 2004. Mol Ecol 16:2958–2973
Prospero S, Grünwald NJ, Winton LM, Hansen EM (2009) Migration patterns of the emerging plant pathogen *Phytophthora ramorum* on the West Coast of the United States of America. Phytopathology 99:739–749
Rademaker JLW, Hoste B, Louws FJ, Kersters K, Swings J, Vauterin L, Vauterin P, De Bruijn FJ (2000) Comparison of AFLP and rep-PCR genomic fingerprinting with DNA-DNA homology studies: *Xanthomonas* as a model system. Int J Syst Evol Microbiol 50:665–677
Rademaker JLW, Louws FJ, Schultz MH, Rossbach U, Vauterin L, Swings J, De Bruijn FJ (2005) A comprehensive species to strain taxonomic framework for *Xanthomonas*. Phytopathology 95:1098–1111
Rizzo DM, Garbelotto M, Davidson JM, Slaughter GW (2002) *Phytophthora ramorum* as the cause of extensive mortality of *Quercus spp.* and *Lithocarpus densiflorus* in California. Plant Dis 86:205–214
Robideau GP, De Cock AWAM, Coffey MD, Voglmayr H, Brouwer H, Bala K, Chitty DW, Désaulniers N, Eggertson QA, Gachon CMM, Hu C-H, Kupper FC, Rintoul TL, Sarhan E, Verstappen ECP, Zhang Y, Bonants PJM, Ristaino JB, Lévesque CA (2011) DNA barcoding of oomycetes with cytochrome oxidase subunit I and internal transcribed spacer. Mol Ecol Resour 11:1002–1911
Robin C, Piou D, Feau N, Douzon G, Schenck N, Hansen EM (2011) Root and aerial infections of *Chamaecyparis lawsoniana* by *Phytophthora lateralis*: a new threat for European countries. For Pathol 41(5):417–424
Rodriguez LM, Grajales A, Arrieta-Ortiz ML, Salazar C, Restrepo S, Bernal A (2012) Genomes-based phylogeny of the genus *Xanthomonas*. BMC Microbiol 12:43
Samson R, Nassan-Agha N (1978) Biovars and serovars among 129 strains of *Erwinia chrysanthemi*. Sta Path Veg Phytobact, INRA, Angers, pp 547–553
Samson R, Legendre JB, Christen R, Fischer-le-Le Saux M, Achouak W, Gardan L (2005) Transfer of *Pectobacterium chrysanthemi* (Burkholder *et al.* 1953) Brenner *et al.* 1973 and *Brenneria paradisiaca* to the genus *Dickeya* gen. nov. as *Dickeya chrysanthemi* comb. nov. and *Dickeya paradisiaca* comb. nov. and delineation of four novel species, *Dickeya dadantii* sp. nov., *Dickeya dianthicola* sp. nov., *Dickeya dieffenbachiae* sp. nov. and *Dickeya zeae* sp. nov. Int J Syst Evol Microbiol 55:1415–1427
Sanfaçon H, Iwanami T, Karasev AV, Van der Vlugt R, Wellink J, Wetzel T, Yoshikawa N (2012) Secoviridae. In: King MJA, Adams MJ, Carstens EB, Lefkowitz EJ (eds) Virus taxonomy, ninth report of the international committee on taxonomy of viruses. Elsevier Academic Press, London, pp 881–899
Schaad NW, Postnikova E, Lacy GH, Sechler A, Agarkova I, Stromberg PE, Stromberg VK, Vidaver AK (2006) Amended classification of *Xanthomonad* pathogens on citrus. Syst Appl Microbiol 29:690–695
Schans J, Ceglarska EB, Cheek S, Stancanelli G (2008) The EFSA panel on plant health: accomplishments and challenges for the EU pest risk assessment. J Plant Pathol 90 (2, Supplement), S2:40
Seinhorst JW (1982) The relationship in field experiments between population density of *Globodera rostochiensis* before planting potatoes and yield of potato tubers. Nematologica 28:277–284
Širca S, Stare BG, Plesko IM, Marn MV, Urek G, Javornik B (2007) *Xiphinema rivesi* from Slovenia transmit *Tobacco ringspot virus* and *Tomato ringspot virus* to cucumber bait plants. Plant Dis 91:770–770
Slawiak M, Beckhoven JRCMV, Speksnijder AGCL, Czajkowski R, Grabe G, Wolf JMVD (2009) Biochemical and genetical analysis reveal a new clade of biovar 3 *Dickeya* spp. strains isolated from potato in Europe. Eur J Plant Pathol 125(2):245–261
Stevens FL (1907) List of New York fungi. J Mycol 13:67–72

Sun XA, Stall RE, Jones JB, Cubero J, Gottwald TR, Graham JH, Dixon WN, Schubert TS, Chaloux PH, Stromberg VK, Lacy GH, Sutton BD (2004) Detection and characterization of a new strain of citrus canker bacteria from key Mexican lime and Alemow in South Florida. Plant Dis 88:1179–1188

Tarjan AC (1969) Variation within the *Xiphinema americanum* group (Nematoda: Longidoridae). Nematologica 15:241–252

Tassi F (1900) Novae Micromycetum species descriptae et iconibus illustratae. Bullettino del Laboratorio ed Orto Botanico, Siena, Vol. III, Fasc. III/IV, 116–132

Taylor CE, Brown DJF (1997) Nematode vectors of plant viruses. CAB International, Wallingford, p 286

Tomilin BA (1979) Key to fungi of the genus *Mycosphaerella Johans*. Nanka, Leningrad, pp 200–201

Torchetti EM, Navarro B, Di Serio F (2012) A single polyprobe for detecting simultaneously eight pospiviroids infecting ornamentals and vegetables. J Virol Methods 186(1–2):141–6

Toth IK, Van der Wolf JM, Saddler G, Lojkowska E, Helias V, Pirhonen M, Tsror L, Elphinstone JG (2011) *Dickeya* species: an emerging problem for potato production in Europe. Plant Pathol 60:385–399

Vaghefi N, Pethybridge SJ, Ford R, Nicolas ME, Crous PW, Taylor PWJ (2012) *Stagonosporopsis* spp. associated with ray blight disease of Asteraceae. Aust Plant Pathol 41:675–686

Van der Aa HA, Noordeloos ME, Gruyter J (1990) Species concept in some larger general coelomycetes. Stud Mycol 32:3–19

Van der Helm F (2009) Bacterierot in Sedum: hoe te handelen? Vakblad voor de Bloemisterij, pp 33–43

Van Doorn J, Vreeburg P, Van Leeuwen P, Dees R (2008) The presence and survival of soft rot (*Erwinia*) in flower bulb production systems. In: X International Symposium on Flower Bulbs and Herbaceous Perennials 886:365–379.

Van Leeuwen P, Dees R, Vreeburg P, En van Doorn J (2012) Oorzaak *Erwinia* problemen dahlia vooral *Dickeya dianthicola*. BloembollenVisie 246:22–23

Vauterin L, Swings J (1997) Are classification and phytopathological diversity compatible in *Xanthomonas* ? J Ind Microbiol Biotechnol 19:77–82

Vauterin L, Hoste B, Kersters K, Swings J (1995) Reclassification of *Xanthomonas*. Int J Syst Bacteriol 45:472–489

Vercauteren A, De Dobbelaere I, Grünwald NJ, Bonants P, Van Bockstaele E, Maes M, Heungens K (2010) Clonal expansion of the Belgian *Phytophthora ramorum* populations based on new microsatellite markers. Mol Ecol 19:92–107

Vercauteren A, Boutet X, D'hondt L, Van Bockstaele E, Maes M, Leus L, Chandelier A, Heungens K (2011) Aberrant genome size and instability of *Phytophthora ramorum* oospore progenies. Fungal Genet Biol 48:537–543

Vernière C, Hartung JS, Pruvost OP, Civerolo EL, Alvarez AM, Maestri P, Luisetti J (1998) Characterization of phenotypically distinct strains of *Xanthomonas axonopodis* pv. citri from Southwest Asia. Eur J Plant Pathol 104:477–487

Vives MC, Velázquez K, Pina JA, Moreno P, Guerri J, Navarro L (2013) Identification of a new Enamovirus associated with Citrus Vein Enation disease by deep sequencing of small RNAs. Phytopathology 103(10):1077–1086

Walker J, Baker K (1983) The correct binomial for the chrysanthemum ray blight pathogen in relation to its geographical distribution. Trans Br Mycol Soc 80:31–38

Webster CG, Reitz SR, Perry KL, Adkins S (2011) A natural M RNA reassortant arising from two species of plant- and insect-infecting bunyaviruses and comparison of its sequence and biological properties to parental species. Virology 413:216–225

Werres S, Marwitz R, Man In T, Veld WA, De Cock AWAM, Bonants PJM, De Weerdt M, Themann K, Ilieva E, Baayen RP (2001) *Phytophthora ramorum* sp. nov., a new pathogen on *Rhododendron* and *Viburnum*. Mycol Res 105:1155–1165

Whitfield AE, Ullman DE, German TL (2005) Tospovirus-thrips interactions. Annu Rev Phytopathol 43:459–489

Wilcox WF, Scott PH, Hamm PB, Kennedy DM, Duncan JM, Brasier CM, Hansen EM (1993) Identity of a *Phytophthora* species attacking raspberry in Europe and North America. Mycol Res 97(7):817–831, 33

Young JM, Bradbury JF, Gardan L, Gvozdyak RI, Stead DE, Takikawa Y, Vidaver AK (1991) Comment on the reinstatement of *Xanthomonas citri* (ex Hasse 1915) Gabriel *et al.* 1989 and *X. phaseoli* (ex Smith 1897) Gabriel *et al.* 1989: Indication of the need for minimal standards for the genus *Xanthomonas*. Int J Syst Evol Microbiol 41:172–177

Young JM, Bull CT, De Boer SH, Firrao G, Gardan L, Saddler GE, Stead DE, Takikawa Y (2001) Classification, nomenclature and plant pathogenic bacteria-A clarification. Phytopathology 91:617–620

Young JM, Park DC, Shearman HM, Fargier E (2008) A multilocus sequence analysis of the genus *Xanthomonas*. Syst Appl Microbiol 31:366–377

Index

M.L. Gullino and P.J.M. Bonants (eds.), *Detection and Diagnostics of Plant Pathogens*, Plant Pathology in the 21st Century 5, DOI 10.1007/978-94-017-9020-8,

T

U

V

MIX
Papier aus verantwortungsvollen Quellen
Paper from responsible sources
FSC® C105338

If you have any concerns about our products,
you can contact us on
ProductSafety@springernature.com

In case Publisher is established outside the EU,
the EU authorized representative is:
Springer Nature Customer Service Center GmbH
Europaplatz 3, 69115 Heidelberg, Germany

Printed by Libri Plureos GmbH
in Hamburg, Germany